MINI-EXAMS

for the

Engineer-In-Training Examination

Michael R. Lindeburg, P.E.

PROFESSIONAL PUBLICATIONS, INC.
Belmont, CA 94002

In the ENGINEERING REVIEW MANUAL SERIES

Engineer-In-Training Review Manual
    Engineering Fundamentals Quick Reference Cards
    Mini-Exams for the E-I-T Exam
    1001 Solved Engineering Fundamentals Problems
    E-I-T Review: A Study Guide
Civil Engineering Reference Manual
    Civil Engineering Quick Reference Cards
    Civil Engineering Sample Examination
    Civil Engineering Review Course on Cassettes
    Seismic Design for the Civil P.E. Exam
    Timber Design for the Civil P.E. Exam
Structural Engineering Practice Problem Manual
Mechanical Engineering Reference Manual
    Mechanical Engineering Quick Reference Cards
    Mechanical Engineering Sample Examination
    101 Solved Mechanical Engineering Problems
    Mechanical Engineering Review Course on Cassettes
    Consolidated Gas Dynamics Tables
Electrical Engineering Reference Manual
Chemical Engineering Reference Manual
    Chemical Engineering Practice Exam Set
Land Surveyor Reference Manual
Metallurgical Engineering Practice Problem Manual
Petroleum Engineering Practice Problem Manual
Expanded Interest Tables
Engineering Law, Design Liability, and Professional Ethics
Engineering Unit Conversions

In the ENGINEERING CAREER ADVANCEMENT SERIES

How to Become a Professional Engineer
The Expert Witness Handbook—A Guide for Engineers
Getting Started as a Consulting Engineer
Intellectual Property Protection—A Guide for Engineers
E-I-T/P.E. Course Coordinator's Handbook
Becoming a Professional Engineer

Distributed by: Professional Publications, Inc.
                  1250 Fifth Avenue
                  Department 77
                  Belmont, CA 94002
                  (415) 593-9119

MINI-EXAMS
for the Engineer-In-Training Examination

Printed in the United States of America

ISBN: 0-932276-34-2

Professional Publications, Inc.
1250 Fifth Avenue, Belmont, CA 94002

Current printing of this edition (last number): 15 14   13   12   11   10

TABLE OF CONTENTS

How to Use these MINI-EXAMS

These *MINI-EXAMS* are an essential part of your preparation for the Engineer-In-Training examination.

Unlike the *SAMPLE EXAMINATION*, which should be taken after your preparation is completed, these *MINI-EXAMS* should be taken as you go along. There is a *MINI-EXAM* for every examination subject. You should take a *MINI-EXAM* each time you finish studying a subject.

The *MINI-EXAMS* contain 15 multiple-choice problems in the same format as Part A of the Engineer-In-Training examination. Knowing the answers to the *MINI-EXAM* problems is half of your goal. The other half requires that you be able to finish each *MINI-EXAM* in 30 minutes or less since speed is critical in the examination.

Answers to the *MINI-EXAM* problems are included in this booklet. If you do poorly on a *MINI-EXAM*, or if you take significantly more than 30 minutes to complete it, you should consider additional study in that subject.

Do not look at the *MINI-EXAMS* until you are ready to take them, as that may influence you to study certain topics with undue emphasis.

These *MINI-EXAMS* are meant to be used in conjunction with the *ENGINEER-IN-TRAINING REVIEW MANUAL* and the *QUICK REFERENCE SUMMARY CARDS*. Do not attempt to use these *MINI-EXAMS* independently or as your sole means of preparation for the exam.

ENGINEER-IN-TRAINING
Mini-Exam

Mathematics

NOTICE TO EXAMINEE: This *MINI-EXAM* has been prepared to familiarize you with the types of questions that could appear on the Engineer-In-Training examination. If this *MINI-EXAM* is to be used as a measure of your speed and preparedness, it must <u>not</u> be taken prior to the completion of your review of the subject.

2

INSTRUCTIONS TO EXAMINEE: Please begin by entering your name, address, *MINI-EXAM* subject, and student number (if applicable) below.

All questions in this *MINI-EXAM* are multiple choice. After solving a problem, choose the closest answer given. Do not expect exact answers. Record your choice by blackening the appropriate circle below.

If it is your intent to take this *MINI-EXAM* under actual test conditions, you should allow yourself 30 minutes to complete all 15 questions. The time available per question under this time limitation is approximately the same as it will be in the actual Engineer-In-Training examination.

After you have completed the entire *MINI-EXAM*, turn to the solutions at the end of this booklet. Page, table, and figure numbers in the solutions refer to the *ENGINEER-IN-TRAINING REVIEW MANUAL*, 6th edition, by Michael R. Lindeburg.

MINI-EXAM subject: _____

1 Ⓐ Ⓑ Ⓒ Ⓓ Ⓔ   6 Ⓐ Ⓑ Ⓒ Ⓓ Ⓔ   11 Ⓐ Ⓑ Ⓒ Ⓓ Ⓔ
2 Ⓐ Ⓑ Ⓒ Ⓓ Ⓔ   7 Ⓐ Ⓑ Ⓒ Ⓓ Ⓔ   12 Ⓐ Ⓑ Ⓒ Ⓓ Ⓔ
3 Ⓐ Ⓑ Ⓒ Ⓓ Ⓔ   8 Ⓐ Ⓑ Ⓒ Ⓓ Ⓔ   13 Ⓐ Ⓑ Ⓒ Ⓓ Ⓔ
4 Ⓐ Ⓑ Ⓒ Ⓓ Ⓔ   9 Ⓐ Ⓑ Ⓒ Ⓓ Ⓔ   14 Ⓐ Ⓑ Ⓒ Ⓓ Ⓔ
5 Ⓐ Ⓑ Ⓒ Ⓓ Ⓔ   10 Ⓐ Ⓑ Ⓒ Ⓓ Ⓔ   15 Ⓐ Ⓑ Ⓒ Ⓓ Ⓔ

Name_____ Student No. _____

Address_____

_____

1. What is the determinant of the following matrix?

$$\begin{bmatrix} 6 & -4 & 2 \\ 9 & 6 & 1 \\ 3 & -1 & -2 \end{bmatrix}$$

(A) +20   (B) -36   (C) -204   (D) +73
(E) -120

2. What is the general solution to the differential equation given below?

$$y'' + 4y' + 4y = 0$$

(A) $y = A_1e^{-2t} - A_2e^{2t}$
(B) $y = A_1e^{-2t} + A_2te^{-2t}$
(C) $y = A_1e^{-2t} + A_2e^{-2t^2}$
(D) $y = Ae^{-2t}$
(E) $y = Ae^{2t}$

3. Evaluate the definite integral below:

$$\int_2^5 (1/x^2)\, dx$$

(A) +.7   (B) +.1   (C) -3   (D) +.3
(E) -.3

4. Evaluate the following limit:

$$\lim_{x \to 4} \frac{x(x^3-6)}{x-5}$$

(A) +232   (B) 0   (C) -48   (D) -232
(E) +48

5. The equation $x^2 + y^2 -4x + 2y - 20 = 0$ describes

(A) a sphere centered at the origin
(B) a circle of radius 5 centered at the origin
(C) an ellipse centered at (2, -1)
(D) a circle of radius 5 centered at (2, -1)
(E) a parabola with vertex at (2, -1)

6. Which of the following is an equation of a line perpendicular to the line $y = 3x + 12$?

(A) $3y = x + 6$
(B) $y = 4x + 1$
(C) $y = -3x + 4$
(D) $9y = -3x + 17$
(E) $8y = 7x + 6$

7. What is the derivative of the following function?

$$z = 1/\sqrt{2k^2 + 8}$$

(A) $4k\sqrt{2k^2 + 8}$
(B) $-2k/\sqrt{(2k^2 + 8)^3}$
(C) $\sqrt{2k^2 + 8}$
(D) $k/\sqrt{2k^2 + 8}$
(E) $-1/2\sqrt[3]{2k^2 + 8}$

8. Which of the following is a relative minimum of

$$f(x) = (1/3)x^3 + \tfrac{1}{2}x^2 - 12x + 34\ ?$$

(A) 68.7   (B) 0   (C) 2.5   (D) 27.3
(E) 11.5

9. Box A has 4 white balls, 3 blue balls, and 3 orange balls. Box B has 2 white balls, 4 blue balls, and 4 orange balls. If one ball is drawn from each box, what is the probability that one of the two balls will be orange?

(A) 7/10   (B) 9/50   (C) 7/25   (D) 27/50
(E) 23/50

10. Which of the following is a solution to the differential equation below?

$$x'' + 4x' -12x = 0$$

(A) $x = 6$
(B) $x = Ae^{-6t}$
(C) $x = A_1 + A_2e^{-6t} - A_3e^{2t}$
(D) $x = Ae^{-2t}$
(E) $x = A_1e^{6x} + A_2e^{-2x}$

11. Determine the value of the following limit:

$$\lim_{x \to 0} \frac{x^2 - 4x + 3}{4x}$$

(A) 3   (B) ∞ (C) ¾   (D) 0
(E) -1

12. How many ways can you invite one or more of five friends to a party?

(A) 31   (B) 63   (C) 25   (D) 48
(E) 120

4

13. What is the area between $y = 0$, $y = 3x^2$, $x = 0$, and $x = 2$?

(A) 12   (B) 6   (C) 3   (D) 24
(E) 8

14. The absolute minimum for $f(x) = x^2/(1 + x^2)$ in the interval $9 \leqslant x \leqslant 10$ is

(A) 100/101   (B) 0   (C) 81/82   (D) .0027
(E) 3

15.   The average number of customers served at a post office is 20 per hour. What is the probability that exactly 8 customers will be served in the next hour?

(A) .0013   (B) .160   (C) .40   (D) .0063
(E) .070

ENGINEER-IN-TRAINING
Mini-Exam

Engineering Economic Analysis

NOTICE TO EXAMINEE: This *MINI-EXAM* has been prepared to familiarize you with the types of questions that could appear on the Engineer-In-Training examination. If this *MINI-EXAM* is to be used as a measure of your speed and preparedness, it must <u>not</u> be taken prior to the completion of your review of the subject.

6

INSTRUCTIONS TO EXAMINEE: Please begin by entering your name, address, *MINI-EXAM* subject, and student number (if applicable) below.

All questions in this *MINI-EXAM* are multiple choice. After solving a problem, choose the closest answer given. Do not expect exact answers. Record your choice by blackening the appropriate circle below.

If it is your intent to take this *MINI-EXAM* under actual test conditions, you should allow yourself 30 minutes to complete all 15 questions. The time available per question under this time limitation is approximately the same as it will be in the actual Engineer-In-Training examination.

After you have completed the entire *MINI-EXAM*, turn to the solutions at the end of this booklet. Page, table, and figure numbers in the solutions refer to the *ENGINEER-IN-TRAINING REVIEW MANUAL*, 6th edition, by Michael R. Lindeburg.

MINI-EXAM subject: _____

1 (A) (B) (C) (D) (E)   6 (A) (B) (C) (D) (E)   11 (A) (B) (C) (D) (E)
2 (A) (B) (C) (D) (E)   7 (A) (B) (C) (D) (E)   12 (A) (B) (C) (D) (E)
3 (A) (B) (C) (D) (E)   8 (A) (B) (C) (D) (E)   13 (A) (B) (C) (D) (E)
4 (A) (B) (C) (D) (E)   9 (A) (B) (C) (D) (E)   14 (A) (B) (C) (D) (E)
5 (A) (B) (C) (D) (E)   10 (A) (B) (C) (D) (E)   15 (A) (B) (C) (D) (E)

Name_____ Student No. _____

Address_____

_____

1. The fixed costs of a company are $500,000 per year. Variable costs are 50% of sales. If the annual sales amount to $750,000, what is the annual profit or loss?

(A) $125,000 loss   (B) break-even
(C) $250,000 profit   (D) $125,000 profit
(E) $50,000 loss

---

2. What is the present worth of $27,000 due in 6 years if money is worth 13% and is compounded semi-annually?

(A) $23,752   (B) $6,229   (C) $12,681
(D) $3,781   (E) $30,510

---

3. A machine costs $6,600. The accounting life of the machine is estimated to be 10 years. If the value of the machine after 10 years is $1,600, what is the depreciation in the fifth year? Use sum-of-the-years' digits method.

(A) $454.54   (B) $500.00   (C) $654.54
(D) $2,776.21   (E) $545.45

---

4. An interest rate is quoted at being 7½% compounded quarterly. What is the effective annual interest rate?

(A) 33.56%   (B) 9.18%   (C) 7.71%
(D) 14.63%   (E) 7.45%

---

5. An annuity of $5,600 is paid each year for 10 years. The payment is made at year-end. If the interest rate is 10%, what is the present worth of the 10 payments?

(A) $14,526   (B) $89,247   (C) $911
(D) $56,000   (E) $34,410

---

6. Building type A has a life of 10 years. The initial cost and cost of replacement every 10 years is $10,000. The annual maintenance is $100. $500 worth of repairs are expected after the first 5 years of each building's life. What is the capitalized cost of a type A building? Assume an interest rate of 8%

(A) $10,000   (B) $20,500   (C) $21,500
(D) $11,250   (E) $21,800

---

7. You can either pay $7000 now for an asset or pay $1200 down and $900 at the end of each year for 10 years. What is the effective annual interest rate for the time payment series?

(A) 6%   (B) 8%   (C) 9%
(D) 10%   (E) 13%

---

8. A perpetual-life project calls for investing $200,000 now and an equal amount 5 years from now. What is the capitalized cost of the project at 6%?

(A) $8,300   (B) $59,000   (C) $791,000
(D) $350,000   (E) $5,820,000

---

9. A $15,000 drill press will be depreciated over 10 years by the straight-line method. If the salvage value is $2,800, what percentage of the initial book value is the annual depreciation?

(A) 8%   (B) 10%   (C) 12%
(D) 14%   (E) 16%

---

10. If a new tool shed costs $1000, has a salvage value of $200 at the end of 8 years, requires annual maintenance of $15, and the minimum return on investment is 20%, what is the present worth?

(A) -47   (B) -58   (C) -1000
(D) -1010   (E) -1100

---

11. What is the maximum profit when the profit-versus-production function is as given below? P is profit and X is units of production.

$$P = 200,000 - .1X - [1.1/(X+1)]^8$$

(A) 200,000   (B) 235,000   (C) 287,000
(D) 305,000   (E) 1,574,000

---

12. An investment costs $8,000 and will have a 10% salvage value in 5 years. Maintenance is $1,000 per year, and annual property taxes are 1% of the original cost. What is the present worth of all costs? The interest rate is 10%.

(A) -3,600   (B) -8,000   (C) -11,300
(D) -11,600   (E) -12,600

---

13. A factory is running at 80% of capacity and 100% efficiency. Fixed costs are $200,000 per period. Variable cost per unit is $1.53. Selling price per unit is $5.75. The production capacity per period is 1,000,000 units. What is the current profit or loss (in thousands of dollars) per period if the inventory remains unchaged?

(A) -4,020 (loss)   (B) 3,180   (C) 3,220
(D) 4,020   (E) 4,420

---

14. Equations for the production costs of two items are given below. What is the cost at the break-even point?

$$C_1 = 60,000 + .021X$$
$$C_2 = 78,000 + .008X$$

(A) 90,000   (B) 890,000   (C) 1,384,600
(D) 2,300,000   (E) 2,900,000

---

15. What is the expected maintenance cost for a machine whose possible maintenance cost has the following distribution? p[$200] = .2; p[$300] = .3; p[$400] = .3; p[$500] = .1; p[$600] = .1.

(A) $1,000   (B) $575   (C) $400
(D) $395   (E) $360

ENGINEER-IN-TRAINING Mini-Exam

Fluid Statics

NOTICE TO EXAMINEE: This *MINI-EXAM* has been prepared to familiarize you with the types of questions that could appear on the Engineer-In-Training examination. If this *MINI-EXAM* is to be used as a measure of your speed and preparedness, it must <u>not</u> be taken prior to the completion of your review of the subject.

10

INSTRUCTIONS TO EXAMINEE: Please begin by entering your name, address, *MINI-EXAM* subject, and student number (if applicable) below.

All questions in this *MINI-EXAM* are multiple choice. After solving a problem, choose the closest answer given. Do not expect exact answers. Record your choice by blackening the appropriate circle below.

If it is your intent to take this *MINI-EXAM* under actual test conditions, you should allow yourself 30 minutes to complete all 15 questions. The time available per question under this time limitation is approximately the same as it will be in the actual Engineer-In-Training examination.

After you have completed the entire *MINI-EXAM*, turn to the solutions at the end of this booklet. Page, table, and figure numbers in the solutions refer to the *ENGINEER-IN-TRAINING REVIEW MANUAL*, 6th edition, by Michael R. Lindeburg.

MINI-EXAM subject: _____

1 Ⓐ Ⓑ Ⓒ Ⓓ Ⓔ   6 Ⓐ Ⓑ Ⓒ Ⓓ Ⓔ   11 Ⓐ Ⓑ Ⓒ Ⓓ Ⓔ
2 Ⓐ Ⓑ Ⓒ Ⓓ Ⓔ   7 Ⓐ Ⓑ Ⓒ Ⓓ Ⓔ   12 Ⓐ Ⓑ Ⓒ Ⓓ Ⓔ
3 Ⓐ Ⓑ Ⓒ Ⓓ Ⓔ   8 Ⓐ Ⓑ Ⓒ Ⓓ Ⓔ   13 Ⓐ Ⓑ Ⓒ Ⓓ Ⓔ
4 Ⓐ Ⓑ Ⓒ Ⓓ Ⓔ   9 Ⓐ Ⓑ Ⓒ Ⓓ Ⓔ   14 Ⓐ Ⓑ Ⓒ Ⓓ Ⓔ
5 Ⓐ Ⓑ Ⓒ Ⓓ Ⓔ   10 Ⓐ Ⓑ Ⓒ Ⓓ Ⓔ   15 Ⓐ Ⓑ Ⓒ Ⓓ Ⓔ

Name_____ Student No. _____

Address_____

_____

1. Absolute viscosity is essentially independent of pressure and is primarily dependent on

(A) specific gravity
(B) density
(C) velocity
(D) pipe wall smoothness
(E) temperature

2. What is the total force on the 8' x 8' square gate shown below? The fluid is water.

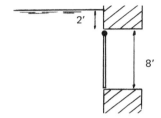

(A) 62 lb
(B) 375 lb
(C) 170 lb
(D) 24,000 lb
(E) 40,000 lb

3. A submerged body will be stable if the center of buoyancy

(A) is located at the body's centroid.
(B) and center of gravity are in the same horizontal plane.
(C) coincides with the center of gravity.
(D) is directly below the center of gravity.
(E) is directly above the center of gravity.

4. An object weighs 65 pounds in air and 42 pounds in water. What is its approximate specific gravity?

(A) 1.0
(B) 1.6
(C) 1.5
(D) 2.8
(E) 16.2

5. A ship floating in sea water (density = 64.0 lb/ft^3) displaces 4500 cubic feet. What is the approximate weight of the ship?

(A) 4500 tons
(B) 200 tons
(C) 40 tons
(D) 60 tons
(E) 144 tons

6. A gauge on a steam condenser reads 24 inches of mercury. If atmospheric pressure is 14.7 psia, what is the pressure inside the condenser?

(A) 1400 psfa
(B) 5000 psfa
(C) 7 psfa
(D) 420 psfa
(E) 620 psfa

7. A sphere 4 inches in diameter just closes a hole (4 inches in diameter) in the bottom of a container. If water in the container rises 8 inches above the container bottom, what is the vertical force on the hemispherical surface?

(A) 3 lb
(B) 6 lb
(C) 1.5 lb
(D) 9 lb
(E) 12 lb

8. The kinematic viscosity of an oil is 0.20 ft^2/sec. Its specific gravity is 0.87. What is its dynamic viscosity?

(A) .34 slugs/ft-sec
(B) .48 slugs/ft-sec
(C) 1.12 slugs/ft-sec
(D) .07 slugs/ft-sec
(E) .65 slugs/ft-sec

9. A balloon is stable because

(A) the center of gravity is above the center of buoyancy.
(B) the center of pressure is above the center of gravity.
(C) the center of buoyancy is above the center of gravity.
(D) the center of buoyancy is above the center of pressure.
(E) the center of gravity coincides with the center of buoyancy.

10. The term g_c which frequently appears in fluid problems has the units of

(A) no units
(B) ft/sec
(C) ft/sec^2
(D) sec^2/ft
(E) lbm-ft/(lbf-sec^2)

11. A rectangular gate (4' wide and 6' high) is submerged vertically as shown below. The fluid is water. What is the total force on the gate?

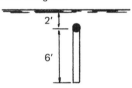

(A) 7490 pounds
(B) 2000 pounds
(C) 12000 pounds
(D) 1250 pounds
(E) 1880 pounds

12. Referring to the gate in problem 11, how many feet below the water surface does the resultant act?

(A) 7.2
(B) 7.6
(C) 3.6
(D) 5.0
(E) 5.6

13. A steel pipe has an allowable stress of 18,000 psi in tension. The pipe is 48" in diameter. The pressure inside the pipe corresponds to a head of 300 feet of water. What is the minimum wall thickness required?

(A) .08"
(B) .14"
(C) .17"
(D) .23"
(E) .37"

14. For a cylindrical tank under pressure, which of the following statements is true?

(A) The circumferential stress is equal to the longitudinal stress.
(B) The longitudinal stress is higher than the circumferential stress.
(C) The circumferential and longitudinal stresses must be added algebraically to obtain the maximum stress.
(D) The circumferential and longitudinal stresses must be added vectorially to obtain the maximum stress.
(E) None of the above are true.

15. On a day when the barometric pressure is 29.92 inches of mercury, a vacuum gage indicates 20 inches of mercury. What is the approximate absolute pressure corresponding to the gage reading?

(A) 5 psi   (B) 19 psi   (C) 10 psi
(D) 25 psi   (E) 13 psi

ENGINEER-IN-TRAINING
Mini-Exam

Fluid Dynamics

NOTICE TO EXAMINEE: This *MINI-EXAM* has been prepared to familiarize you with the types of questions that could appear on the Engineer-In-Training examination. If this *MINI-EXAM* is to be used as a measure of your speed and preparedness, it must <u>not</u> be taken prior to the completion of your review of the subject.

14

INSTRUCTIONS TO EXAMINEE: Please begin by entering your name, address, *MINI-EXAM* subject, and student number (if applicable) below.

All questions in this *MINI-EXAM* are multiple choice. After solving a problem, choose the closest answer given. Do not expect exact answers. Record your choice by blackening the appropriate circle below.

If it is your intent to take this *MINI-EXAM* under actual test conditions, you should allow yourself 30 minutes to complete all 15 questions. The time available per question under this time limitation is approximately the same as it will be in the actual Engineer-In-Training examination.

After you have completed the entire *MINI-EXAM*, turn to the solutions at the end of this booklet. Page, table, and figure numbers in the solutions refer to the *ENGINEER-IN-TRAINING REVIEW MANUAL*, 6th edition, by Michael R. Lindeburg.

MINI-EXAM subject: _____

1 Ⓐ Ⓑ Ⓒ Ⓓ Ⓔ   6 Ⓐ Ⓑ Ⓒ Ⓓ Ⓔ   11 Ⓐ Ⓑ Ⓒ Ⓓ Ⓔ
2 Ⓐ Ⓑ Ⓒ Ⓓ Ⓔ   7 Ⓐ Ⓑ Ⓒ Ⓓ Ⓔ   12 Ⓐ Ⓑ Ⓒ Ⓓ Ⓔ
3 Ⓐ Ⓑ Ⓒ Ⓓ Ⓔ   8 Ⓐ Ⓑ Ⓒ Ⓓ Ⓔ   13 Ⓐ Ⓑ Ⓒ Ⓓ Ⓔ
4 Ⓐ Ⓑ Ⓒ Ⓓ Ⓔ   9 Ⓐ Ⓑ Ⓒ Ⓓ Ⓔ   14 Ⓐ Ⓑ Ⓒ Ⓓ Ⓔ
5 Ⓐ Ⓑ Ⓒ Ⓓ Ⓔ   10 Ⓐ Ⓑ Ⓒ Ⓓ Ⓔ   15 Ⓐ Ⓑ Ⓒ Ⓓ Ⓔ

Name_____ Student No. _____

Address_____

_____

15

1. Reynolds number may be calculated from:

(A) diameter, velocity, and absolute viscosity
(B) diameter, velocity, and surface tension
(C) diameter, density, and kinematic viscosity
(D) diameter, density, and absolute viscosity
(E) characteristic length, mass flow rate per unit area, and absolute viscosity

2. Roughening the leading edge of a smooth sphere will reduce its drag coefficient because

(A) the wake width increases
(B) the separation points move to the front of the sphere
(C) the wake eddies increase
(D) the boundary layer becomes turbulent
(E) Stoke's law becomes applicable

3. What is the hydraulic radius of a rectangular flume 2 feet high and 4 feet wide which is running half full?

(A) 1.33 feet
(B) .33
(C) 8.0
(D) .40
(E) .67

4. Water flows at 10 ft/sec in a 1" inside diameter pipe. What is the velocity if the pipe suddenly increases in diameter to 2"?

(A) 5 ft/sec
(B) 2.5
(C) 40
(D) 20
(E) answer depends on the flow direction

5. What pressure differential exists across a perfect venturi with an area reduction ratio of (3:1) if water is flowing through the throat at 40 fps?

(A) .6 feet of water
(B) 17
(C) 22
(D) 27
(E) 1378

6. If 'L' is defined as the characteristic length, what does the quantity (v^2/Lg) represent?

(A) velocity pressure
(B) Reynolds number
(C) Froude number
(D) total pressure
(E) static pressure

7. Minor losses through valves, fittings, diameter changes, and bends are proportional to

(A) total head
(B) dynamic head
(C) static head
(D) wet head
(E) velocity

8. The horsepower of an ideal pump used to move 2 cfs of water into a tank 50 feet above the pump is most nearly

(A) 2
(B) 11
(C) 290
(D) 1213
(E) 6240

9. A horizontal pipe section 1000 feet long has a total energy loss of 26.2 feet. If the inside pipe diameter is 12 inches and the flow velocity is 10 ft/sec, what is the Darcy-Weisbach friction coefficient?

(A) 0.0170
(B) 0.0080
(C) 0.0017
(D) 0.0002
(E) 0.0008

10. The Reynolds number for a 1-foot diameter sphere moving through a fluid (specific gravity of 1.22, absolute viscosity of 0.00122 lb-sec/ft^2) at 10 ft/sec is approximately

(A) 20
(B) 200
(C) 2,000
(D) 20,000
(E) 200,000

11. Water is flowing in a circular pipe between points 1 and 2. The pressure at point 1 is 16.8 psia. The pressure and velocity at point 2 are 17.2 psia and 6.2 ft/sec, respectively. Points 1 and 2 are at the same elevation. Neglecting friction, what is the velocity at point 1?

(A) 97.8 ft/sec
(B) 9.9
(C) 21.0
(D) 1.52
(E) 4.58

12. The critical depth in a rectangular channel 8 feet wide flowing at a critical velocity of 2 ft/sec is approximately

(A) 0.12 feet
(B) 2.00
(C) 4.00
(D) 0.06
(E) 0.08

13. At a certain section of pipe, water is flowing at a pressure of 80 psi and with a linear velocity of 9 ft/sec. What is the total flow work for 1.5 cubic feet of water which pass that section?

(A) 18,000 ft-lb
(B) 36,000
(C) 120
(D) 12,000
(E) 0

14. A blower type fan delivers 1200 cfm when turning at 720 rpm. The speed is increased to 750 rpm and the fan delivers 1250 cfm against the same static pressure. What is the percentage increase in required power?

(A) 10% more power
(B) 21 %
(C) 33%
(D) 300%
(E) no increase

15. A jet of water with a cross sectional area of one square inch flows at the rate of 0.139 cfs. The jet impinges on a stationary blade and is deflected 120° from its original path. If the effects of friction are ignored, what is the force on the blade?

(A) 2.69 lb
(B) 4.04
(C) 5.38
(D) 8.08
(E) none of the above

ENGINEER-IN-TRAINING
Mini-Exam

Thermodynamics

NOTICE TO EXAMINEE: This *MINI-EXAM* has been prepared to familiarize you with the types of questions that could appear on the Engineer-In-Training examination. If this *MINI-EXAM* is to be used as a measure of your speed and preparedness, it must <u>not</u> be taken prior to the completion of your review of the subject.

18

INSTRUCTIONS TO EXAMINEE: Please begin by entering your name, address, *MINI-EXAM* subject, and student number (if applicable) below.

All questions in this *MINI-EXAM* are multiple choice. After solving a problem, choose the closest answer given. Do not expect exact answers. Record your choice by blackening the appropriate circle below.

If it is your intent to take this *MINI-EXAM* under actual test conditions, you should allow yourself 30 minutes to complete all 15 questions. The time available per question under this time limitation is approximately the same as it will be in the actual Engineer-In-Training examination.

After you have completed the entire *MINI-EXAM*, turn to the solutions at the end of this booklet. Page, table, and figure numbers in the solutions refer to the *ENGINEER-IN-TRAINING REVIEW MANUAL*, 6th edition, by Michael R. Lindeburg.

MINI-EXAM subject: _____

```
1  A B C D E      6  A B C D E     11  A B C D E
2  A B C D E      7  A B C D E     12  A B C D E
3  A B C D E      8  A B C D E     13  A B C D E
4  A B C D E      9  A B C D E     14  A B C D E
5  A B C D E     10  A B C D E     15  A B C D E
```

Name_____ Student No. _____

Address_____

_____

1. Which of the following statements about constant pressure and constant volume process lines on the T-s diagram is true?

(A) The processes can be depicted by a straight vertical line.
(B) The processes can be depicted by a straight horizontal line.
(C) The area under the curve represents the work done.
(D) The area above the curve represents the work done.
(E) The area under the curve represents the heat transfer.

2. Which of the following relationships is true for an ideal gas going through a reversible adiabatic process?

(A) $c_v dT + pdV = 0$

(B) $c_v dT + Vdp = 0$

(C) $c_p dT + pdV = 0$

(D) $du + dh = ds$

(E) $Tds = pdV$

3. A ball drops onto a solid, massive surface. The surface does not deform. The ball deforms and returns to shape without a change in temperature. Which of the following statements is true?

(A) The ball's enthalpy has increased.
(B) The potential energy of the ball has increased.
(C) The ball's entropy has increased.
(D) The ball's internal energy has increased.
(E) None of the above are true.

4. A perfect gas undergoes a constant 15.3 psig pressure process with a 20 cubic feet change in volume. What is the work required to cause this volume change?

(A) 86,400 BTU/pound
(B) 111
(C) .08
(D) 57
(E) .8

5. Two cubic feet of air are compressed from 14.7 psia to 23.8 psia in an isentropic process. What is the final volume if the air is further compressed to 30 psia in an isentropic process?

(A) 1.2 cubic feet
(B) .6
(C) .37
(D) .74
(E) 1.5

6. How much energy is required to heat 2 pounds of 210°F water to 213°F at a pressure of 14.7 psia?

(A) 6 calories
(B) 3 BTU
(C) 6 BTU
(D) 969 BTU
(E) 1938 BTU

7. An office coffee percolator requires 1 gallon of water each day to replace evaporation losses. The percolator is acting as

(A) a dehumidifier
(B) both a latent and a sensible heat source
(C) a sensible heat source only
(D) a latent heat source only
(E) a hygroscopic device

8. A vapor and a liquid can coexist

(A) only at the triple point
(B) only above the critical pressure
(C) only below the triple point temperature
(D) only above the critical temperature
(E) below the critical temperature

9. Which of the following statements is true of a reversible process?

(A) Heat loss is positive.
(B) The internal energy change and work output are equal.
(C) The decrease in entropy equals the enthalpy loss.
(D) Heat loss is negative.
(E) The increase in entropy equals the enthalpy loss.

10. An ideal gas of molecular weight 24 is contained in a 30 ft^3 tank. If the temperature is 90°F and the pressure is 250 psia inside the tank, what is the weight of the gas in the tank?

(A) .3 pounds
(B) 31
(C) 73
(D) 1213
(E) 573

11. A gas is compressed in a reversible process from 15 psia and 68°F to 75 psia. If the compression process follows a polytropic relationship with n = 1.4, what work is done by the gas during the compression?

(A) 1.7 EE6 ft-lb/pmole
(B) -1.7 EE6
(C) -1.2 EE6
(D) 1.2 EE6
(E) -3.1 EE6

12. Two blocks of copper weighing 50 pounds each are brought into contact. One block is at 100°F and the other is at 500°F. If these two blocks are considered as a system and the surroundings are neglected, determine the change in the entropy of the system. (The specific heat of copper is .0915 BTU/lb-°F.)

(A) 33.6 BTU/°R
(B) -.003
(C) -3.45
(D) .328
(E) 361

13. What is the enthalpy of 100 psia steam with a quality of 97%?

(A) 2049 BTU/pound
(B) 1617
(C) 1182
(D) 195
(E) 1160

14. Air at 400·F and 50 psia is compressed isothermally to one-tenth of its original volume. What is the final pressure?

(A) 100 psia
(B) 1000 psia
(C) 375 psia
(D) 50 psia
(E) 500 psia

15. How much does the increase in enthalpy exceed the increase in internal energy when one pound of ice evaporates to saturated vapor at the triple point?

(A) 76.1 BTU
(B) 26.1
(C) 163
(D) 54.2
(E) 83.7

ENGINEER-IN-TRAINING
Mini-Exam

Power Cycles

NOTICE TO EXAMINEE: This *MINI-EXAM* has been prepared to familiarize you with the types of questions that could appear on the Engineer-In-Training examination. If this *MINI-EXAM* is to be used as a measure of your speed and preparedness, it must <u>not</u> be taken prior to the completion of your review of the subject.

22

INSTRUCTIONS TO EXAMINEE: Please begin by entering your name, address, *MINI-EXAM* subject, and student number (if applicable) below.

All questions in this *MINI-EXAM* are multiple choice. After solving a problem, choose the closest answer given. Do not expect exact answers. Record your choice by blackening the appropriate circle below.

If it is your intent to take this *MINI-EXAM* under actual test conditions, you should allow yourself 30 minutes to complete all 15 questions. The time available per question under this time limitation is approximately the same as it will be in the actual Engineer-In-Training examination.

After you have completed the entire *MINI-EXAM*, turn to the solutions at the end of this booklet. Page, table, and figure numbers in the solutions refer to the *ENGINEER-IN-TRAINING REVIEW MANUAL*, 6th edition, by Michael R. Lindeburg.

MINI-EXAM subject: _____

1 (A) (B) (C) (D) (E)   6 (A) (B) (C) (D) (E)   11 (A) (B) (C) (D) (E)
2 (A) (B) (C) (D) (E)   7 (A) (B) (C) (D) (E)   12 (A) (B) (C) (D) (E)
3 (A) (B) (C) (D) (E)   8 (A) (B) (C) (D) (E)   13 (A) (B) (C) (D) (E)
4 (A) (B) (C) (D) (E)   9 (A) (B) (C) (D) (E)   14 (A) (B) (C) (D) (E)
5 (A) (B) (C) (D) (E)   10 (A) (B) (C) (D) (E)   15 (A) (B) (C) (D) (E)

Name_____ Student No. _____

Address_____

_____

23

1. The ideal Otto cycle consists of

(A) isochoric and isentropic processes
(B) isobaric and isentropic processes
(C) isochoric and isothermal processes
(D) isobaric and isothermal processes
(E) isothermal and isentropic processes

2. What is the efficiency of an ideal Carnot cycle operating between 100°F and 900°F?

(A) 90%   (B) 89%   (C) 59%
(D) 16%   (E) 11%

3. A Carnot refrigeration cycle operates between 15°F and 75°F. What is the coefficient of performance?

(A) 1.3   (B) 3.7   (C) 4.5
(D) 7.9   (E) 8.9

4. Freon-12 is used as the working fluid in an ideal heat pump. The temperature in the condenser and evaporator are 120°F and 10°F respectively. What is the approximate coefficient of performance of this heat pump?

(A) 3.3   (B) 4.3   (C) 5.3
(D) 6.3   (E) 8.0

5. If 18,500 BTU can be produced by burning one pound of a certain fuel, and if an engine has a fuel consumption of 0.40 lb/hp-hr, what is the efficiency of the engine?

(A) 34.4%   (B) 1.7%   (C) 67.37%
(D) 100%   (E) 21.6%

6. A Carnot cycle uses one pound of water as the working fluid. The maximum and minimum temperatures of the cycle are 400°F and 200°F respectively. What is the approximate net work output per cycle?

(A) 100 BTU   (B) 200 BTU   (C) 300 BTU
(D) 400 BTU   (E) 500 BTU

7. The flow rate through a steam turbine is 25,000 pounds per hour. The inlet and outlet steam velocities are 6,000 fpm and 25,000 fpm respectively. The heat loss through the casing is 125,000 BTU/hour. The inlet and outlet steam enthalpies are 1200 BTU/lb and 900 BTU/lb respectively. What is the shaft horsepower?

(A) 20.14   (B) 1715   (C) 3313
(D) 5.27   (E) 2865

8. 20 psia and 40°F air flows steadily through a heater and leaves at 15 psia and 140°F. The heat added is

(A) 6 BTU/lb   (B) 43 BTU/lb   (C) 86 BTU/lb
(D) 24 BTU/lb   (E) 12 BTU/lb

9. Two pounds of air initially at 60 psia and 600°F expand isentropically until the temperature is 200°F. What is the approximate work done during the expansion?

(A) 196 BTU   (B) -135 BTU   (C) 68 BTU
(D) -196 BTU   (E) 135 BTU

10. A steam turbine receives 140 psia steam with an enthalpy of 1192 BTU/lb. The turbine exhausts to 22 psia. The exhaust steam has an enthalpy of 1158 BTU/lb. What is the energy used by the turbine?

(A) 2350 BTU/lb (B) 34 BTU/lb   (C) 42 BTU/lb
(D) 2342 BTU/lb (E) none of the above

11. A new automobile engine has a noticeable knock (ping). Which of the following remedies is the least likely to solve the problem?

(A) Retarding the spark
(B) Switching to a higher octane gasoline
(C) Supercharging the engine
(D) Lowering the compression ratio
(E) Injecting a water/alcohol mixture

12. Which of the following is a unit of power?

(A) hp-hr
(B) kw-hr
(C) ft-lb
(D) BTU/hr
(E) none of the above

13. Cooling the air with intercoolers between stages of a multi-stage air compressor

(A) keeps atmospheric air from condensing.
(B) decreases the volume of air being compressed.
(C) substantially decreases the volumetric efficiency.
(D) decreases the horsepower requirement.
(E) is used in high-temperature climates.

14. Saturated steam at 200 psia is reduced to 50 psia by means of a pressure-reducing (throttling) valve in the steam line. The enthalpy does not change. Which of the following statements is true?

(A) The temperature will remain constant.
(B) The entropy will remain constant.
(C) The specific volume will remain constant.
(D) The steam will be superheated.
(E) The steam will have a high moisture content.

15. Turbojet engine thrust is produced by

(A) The reaction of the high velocity exhaust gases on the still atmospheric air.
(B) The acceleration of the gases through the engine.
(C) The "ram" effect of the intake air.
(D) The additional weight of the injected fuel.
(E) The thrust reaction of the turbine blading.

ENGINEER-IN-TRAINING Mini-Exam

Chemistry

NOTICE TO EXAMINEE: This *MINI-EXAM* has been prepared to familiarize you with the types of questions that could appear on the Engineer-In-Training examination. If this *MINI-EXAM* is to be used as a measure of your speed and preparedness, it must <u>not</u> be taken prior to the completion of your review of the subject.

26

INSTRUCTIONS TO EXAMINEE: Please begin by entering your name, address, *MINI-EXAM* subject, and student number (if applicable) below.

All questions in this *MINI-EXAM* are multiple choice. After solving a problem, choose the closest answer given. Do not expect exact answers. Record your choice by blackening the appropriate circle below.

If it is your intent to take this *MINI-EXAM* under actual test conditions, you should allow yourself 30 minutes to complete all 15 questions. The time available per question under this time limitation is approximately the same as it will be in the actual Engineer-In-Training examination.

After you have completed the entire *MINI-EXAM*, turn to the solutions at the end of this booklet. Page, table, and figure numbers in the solutions refer to the *ENGINEER-IN-TRAINING REVIEW MANUAL*, 6th edition, by Michael R. Lindeburg.

MINI-EXAM subject: _____

1 (A) (B) (C) (D) (E)   6 (A) (B) (C) (D) (E)   11 (A) (B) (C) (D) (E)
2 (A) (B) (C) (D) (E)   7 (A) (B) (C) (D) (E)   12 (A) (B) (C) (D) (E)
3 (A) (B) (C) (D) (E)   8 (A) (B) (C) (D) (E)   13 (A) (B) (C) (D) (E)
4 (A) (B) (C) (D) (E)   9 (A) (B) (C) (D) (E)   14 (A) (B) (C) (D) (E)
5 (A) (B) (C) (D) (E)   10 (A) (B) (C) (D) (E)   15 (A) (B) (C) (D) (E)

Name_____ Student No. _____

Address_____

_____

1. What is the molarity of 2 liters of aqueous solution formed from 588 grams of H_2SO_4?

(A) 1   (B) 1.5   (C) 3
(D) 6   (E) 12

2. Which of the following chemical reactions is not balanced?

(A) $2C_2H_2 + 4O_2 \rightarrow 4CO_2 + 2H_2O$
(B) $3Fe_2O_3 + 9H_2 \rightarrow 6Fe + 9H_2O$
(C) $2Al + Fe_2O_3 \rightarrow Al_2O_3 + 2Fe$
(D) $SiO_2 + 2C \rightarrow Si + 2CO$
(E) $PCl_3 + 3NH_3 \rightarrow P(NH_2)_3 + 3HCl$

3. What is the oxidation number of PO_4 in the compound $Ca_3(PO_4)_2$?

(A) -6   (B) +2   (C) -2
(D) +3   (E) -3

4. Why is less current required to deposit 1 mole of Ag from $AgNO_3$ than 1 mole of Cu from $CuCl_2$ in the same amount of time?

(A) Ag has a higher susceptibility.
(B) $CuCl_2$ is insoluble.
(C) Ag has a lower relaxation constant.
(D) Ag has a smaller oxidation number.
(E) Ag has a lower ionization energy.

5. How many moles of H_3PO_4 are required to neutralize one mole of NaOH?

(A) .33   (B) .5   (C) 1
(D) 2   (E) 3

6. The heats of formation for several compounds are given below. What quantity of heat (in kcal/mole) is liberated when methane is burned completely?

Methane (g): -17.9 kcal/mole
Water (g): -57.8 kcal/mole
CO_2 (g): -94.1 kcal/mole

(A) 12   (B) 3070   (C) 192
(D) 134   (E) none of the above

7. What are the values of a, b, and d when the equation below is balanced?

$$K_2Cr_2O_7 + aHCl \rightarrow bKCl + 2CrCl_3 + cH_2O + dCl_2$$

(A) 14,2,3   (B) 7,1,3   (C) 12,6,1
(D) 14,1,6   (E) 2,6,3

8. What is oxidized and what is reduced in the reaction below?

$$Zn + H_2SO_4 \rightarrow ZnSO_4 + H_2$$

(A) Zinc and hydrogen are both reduced.
(B) Zinc is oxidized and hydrogen is reduced.
(C) Zinc is reduced and hydrogen is oxidized.
(D) Only hydrogen is reduced; nothing is oxidized.
(E) Zinc and hydrogen are both oxidized.

9. If an alkyl radical (C_2H_5) and a hydroxide radical (OH) combine, the result is

(A) Ethanol
(B) Phenol
(C) Ethane
(D) Methanol
(E) Ethene

10. How is the equilibrium constant defined for the following reaction?

$$H_2 + I_2 \rightarrow 2HI$$

(A) $\dfrac{[HI]^2}{[H_2][I_2]}$ (B) $\dfrac{[H][I]}{[H^+][I^-]}$

(C) $\dfrac{[H^+][I^-]}{[HI]^2}$ (D) $\dfrac{[HI]}{[H_2][I_2]}$

(E) $\dfrac{[H_2][I_2]}{[HI]^2}$

11. What is the percent ionization of a (.03)M acetic acid solution? (K = 1.7 EE-5)

(A) 50%   (B) 2.4%   (C) 1.7%
(D) 100%   (E) 24%

12. A compound contains zinc, carbon, and hydrogen. When burned in air, the products are 16.28 grams of ZnO, 35.2 grams of CO_2, and 18 grams of H_2O. What is the empirical formula of this compound?

(A) $ZnCH_3$ (B) ZnC_6H_{12} (C) Zn_2CH_{18}
(D) $Zn_2C_4H_{12}$ (E) ZnC_4H_{10}

13. A buffer solution

(A) may consist of a weak acid and the salt of that acid.
(B) turns an acid red.
(C) magnifies the pH change in a solution.
(D) is used to mix acids.
(E) is the combination of a base and an acid.

28

14. How many grams of $Al_2(SO_4)_3$ are there in 300 ml of (1.50)M solution?

(A) 154   (B) 342   (C) 76
(D) 450   (E) 217

15. If 23.05 ml of 0.1 N NaOH solution is required to neutralize 10.00 ml of an H_2SO_4 solution (of unknown strength), what is the normality of the acid solution?

(A) .1   (B) .27   (C) .85
(D) .23   (E) .56

ENGINEER-IN-TRAINING
Mini-Exam

Statics

NOTICE TO EXAMINEE: This *MINI-EXAM* has been prepared to familiarize you with the types of questions that could appear on the Engineer-In-Training examination. If this *MINI-EXAM* is to be used as a measure of your speed and preparedness, it must <u>not</u> be taken prior to the completion of your review of the subject.

30

INSTRUCTIONS TO EXAMINEE: Please begin by entering your name, address, *MINI-EXAM* subject, and student number (if applicable) below.

All questions in this *MINI-EXAM* are multiple choice. After solving a problem, choose the closest answer given. Do not expect exact answers. Record your choice by blackening the appropriate circle below.

If it is your intent to take this *MINI-EXAM* under actual test conditions, you should allow yourself 30 minutes to complete all 15 questions. The time available per question under this time limitation is approximately the same as it will be in the actual Engineer-In-Training examination.

After you have completed the entire *MINI-EXAM*, turn to the solutions at the end of this booklet. Page, table, and figure numbers in the solutions refer to the *ENGINEER-IN-TRAINING REVIEW MANUAL*, 6th edition, by Michael R. Lindeburg.

MINI-EXAM subject: _____

1 Ⓐ Ⓑ Ⓒ Ⓓ Ⓔ   6 Ⓐ Ⓑ Ⓒ Ⓓ Ⓔ   11 Ⓐ Ⓑ Ⓒ Ⓓ Ⓔ

2 Ⓐ Ⓑ Ⓒ Ⓓ Ⓔ   7 Ⓐ Ⓑ Ⓒ Ⓓ Ⓔ   12 Ⓐ Ⓑ Ⓒ Ⓓ Ⓔ

3 Ⓐ Ⓑ Ⓒ Ⓓ Ⓔ   8 Ⓐ Ⓑ Ⓒ Ⓓ Ⓔ   13 Ⓐ Ⓑ Ⓒ Ⓓ Ⓔ

4 Ⓐ Ⓑ Ⓒ Ⓓ Ⓔ   9 Ⓐ Ⓑ Ⓒ Ⓓ Ⓔ   14 Ⓐ Ⓑ Ⓒ Ⓓ Ⓔ

5 Ⓐ Ⓑ Ⓒ Ⓓ Ⓔ   10 Ⓐ Ⓑ Ⓒ Ⓓ Ⓔ   15 Ⓐ Ⓑ Ⓒ Ⓓ Ⓔ

Name_____ Student No. _____

Address_____

_____

1. What is the y-coordinate of the centroid of the shape shown below?

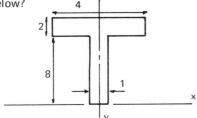

(A) 0   (B) 6.5   (C) 7.2
(D) 7.9   (E) 8.1

2. What is I_y for the shape illustrated in problem 1?

(A) 0   (B) 136   (C) 2.8
(D) 11.3   (E) 821.3

3. What is the radius of gyration about the y axis of the shape illustrated in problem 1?

(A) 51.3   (B) 7.2   (C) .71
(D) .84   (E) .50

4. What is the magnitude of the couple shown below?

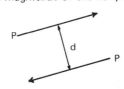

(A) 0   (B) ½Pd   (C) Pd
(D) 2Pd   (E) 1.0

5. Find the distance d such that reactions L and R are equal.

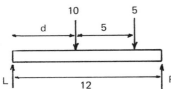

(A) 4.3   (B) 3.7   (C) 7.7
(D) 4.8   (E) none of the above

6. What is the magnitude of force $\vec{F}$?

$$\vec{F} = 6\vec{i} + 2\vec{j} + 10\vec{k}$$

(A) 18.0   (B) 6.1   (C) 3
(D) 11.8   (E) 10.7

7. What force P is required to hold the 1200 pound load in equilibrium using the system of pulleys shown below?

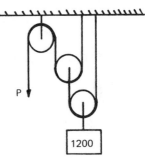

(A) 300 pounds   (B) 200 pounds   (C) 240 pounds
(D) 400 pounds   (E) 600 pounds

8. What is the reaction at A for the simply-supported beam shown below?

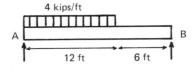

(A) 4,000 lb   (B) 8,000 lb   (C) 32,000 lb
(D) 16,000 lb   (E) 20,000 lb

9. The post in the figure below has a weight of 86 pounds. All surfaces are smooth. What is the tension in cable AB which will keep the post (BC) from sliding?

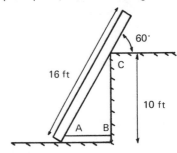

(A) 26.3 lb   (B) 29.8 lb   (C) 27.9 lb
(D) 44.7   (E) 25.8 lb

32

10. Three forces act on the wall bracket as shown below. What is the magnitude of the resultant force?

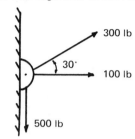

300 lb

30°

100 lb

500 lb

(A) 123 lb   (B) 502 lb   (C) 350 lb
(D) 743 lb   (E) 360 lb

11. What is the horizontal force F that must be applied to the 50 pound weight in order that the cord make a 20° angle with the vertical?

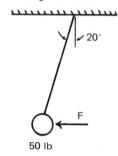

20°

F

50 lb

(A) 18.2 lb   (B) 7.6 lb   (C) 50 lb
(D) 21.0 lb   (E) 35.0 lb

12. The coefficient of static friction between the two blocks and between block B and the wall is 0.25. Block A is fastened to the wall with a cord. What minimum force will cause block B to move?

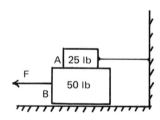

A 25 lb

F

50 lb

B

(A) 10 lb   (B) 25 lb   (C) 50 lb
(D) 100 lb   (E) 7 lb

13. A 400 pound force acts as shown on a 800 pound block. The coefficients of friction are .25 (static) and .20 (dynamic). What is the frictional force acting on the block?

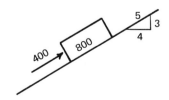

5
3
4

400   800

(A) 160 lb↙   (B) 128 lb↗   (C) 128↙
(D) 80 lb↗   (E) 160 lb↗

14. If the weight of the members is neglected, what is the reaction at A?

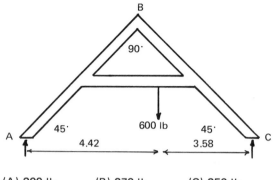

B

90°

A   45°   600 lb   45°   C

4.42   3.58

(A) 200 lb   (B) 270 lb   (C) 250 lb
(D) 170 lb   (E) 130 lb

15. What is the approximate force in cord AB?

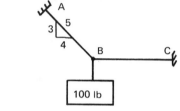

A

3   5
4   B   C

100 lb

(A) 60 lb   (B) 130 lb   (C) 170 lb
(D) 210 lb   (E) 100 lb

ENGINEER-IN-TRAINING
Mini-Exam

Dynamics

NOTICE TO EXAMINEE: This *MINI-EXAM* has been prepared to familiarize you with the types of questions that could appear on the Engineer-In-Training examination. If this *MINI-EXAM* is to be used as a measure of your speed and preparedness, it must <u>not</u> be taken prior to the completion of your review of the subject.

34

INSTRUCTIONS TO EXAMINEE: Please begin by entering your name, address, *MINI-EXAM* subject, and student number (if applicable) below.

All questions in this *MINI-EXAM* are multiple choice. After solving a problem, choose the closest answer given. Do not expect exact answers. Record your choice by blackening the appropriate circle below.

If it is your intent to take this *MINI-EXAM* under actual test conditions, you should allow yourself 30 minutes to complete all 15 questions. The time available per question under this time limitation is approximately the same as it will be in the actual Engineer-In-Training examination.

After you have completed the entire *MINI-EXAM*, turn to the solutions at the end of this booklet. Page, table, and figure numbers in the solutions refer to the *ENGINEER-IN-TRAINING REVIEW MANUAL*, 6th edition, by Michael R. Lindeburg.

MINI-EXAM subject: _____

1 Ⓐ Ⓑ Ⓒ Ⓓ Ⓔ	6 Ⓐ Ⓑ Ⓒ Ⓓ Ⓔ	11 Ⓐ Ⓑ Ⓒ Ⓓ Ⓔ	
2 Ⓐ Ⓑ Ⓒ Ⓓ Ⓔ	7 Ⓐ Ⓑ Ⓒ Ⓓ Ⓔ	12 Ⓐ Ⓑ Ⓒ Ⓓ Ⓔ	
3 Ⓐ Ⓑ Ⓒ Ⓓ Ⓔ	8 Ⓐ Ⓑ Ⓒ Ⓓ Ⓔ	13 Ⓐ Ⓑ Ⓒ Ⓓ Ⓔ	
4 Ⓐ Ⓑ Ⓒ Ⓓ Ⓔ	9 Ⓐ Ⓑ Ⓒ Ⓓ Ⓔ	14 Ⓐ Ⓑ Ⓒ Ⓓ Ⓔ	
5 Ⓐ Ⓑ Ⓒ Ⓓ Ⓔ	10 Ⓐ Ⓑ Ⓒ Ⓓ Ⓔ	15 Ⓐ Ⓑ Ⓒ Ⓓ Ⓔ	

Name_____ Student No. _____

Address_____

_____

1. A 1-slug wheel with a 2-foot radius rotates at 8 radians per second. What is the tangential force required to stop rotation in 5 seconds?

(A) .8 lb   (B) 1.6 lb   (C) 4.0 lb
(D) 8.0 lb   (E) 10.0 lb

---

2. The coefficient of friction between a 100 pound block of ice and the surface on which it moves is .20. What acceleration is experienced by the block due to the 100 pound force shown below?

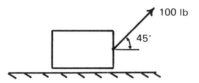

100 lb

45°

(A) .68 ft/sec^2 (B) 13.6 ft/sec^2 (C) 17.1 ft/sec^2
(D) 20.9 ft/sec^2 (E) 25.3 ft/sec^2

---

3. What is the work done against friction when a 10 pound block starts from rest and reaches 52 fps after sliding 100 feet down a 30° incline?

(A) 80 ft-lb   (B) 340 ft-lb   (C) 420 ft-lb
(D) 500 ft-lb   (E) 840 ft-lb

---

4. A linearly-increasing force starts from zero at t=0 and reaches 4 pounds at t=4. The force then decreases linearly and reaches zero at t=6. What is the average force during the interval t = [0,6] ?

(A) 1 lb   (B) 2 lb   (C) 3 lb
(D) 4 lb   (E) 6 lb

---

5. A particle experiences the acceleration shown by the graph below. What is the distance traveled during the time interval t = [0,3] ?

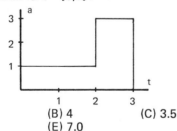

(A) 2   (B) 4   (C) 3.5
(D) 5.5   (E) 7.0

---

6. What force is required to initiate motion of the block shown below if the coefficients of static and dynamic friction are .3 and .2 respectively?

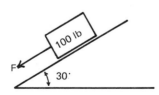

100 lb

F

30°

(A) 0 lb   (B) 10 lb   (C) 20 lb
(D) 30 lb   (E) 40 lb

---

7. If r is the radius of the earth and g is the gravitational acceleration at the earth's surface, what is the gravitational acceleration at altitude h above the earth's surface?

(A) g/h   (B) g/h^2 (C) g/(r+h)
(D) g/(r+h)^2 (E) (gr-2h)/r

---

8. A 5 pound sphere moves down a frictionless plane with a vertical drop of 1 for every horizontal distance of 4. What is the velocity after the sphere has experienced a vertical displacement of 1 foot?

(A) 64.4 fps   (B) 8.0 fps   (C) 32.2 fps
(D) 33.1 fps   (E) 1094.0 fps

---

9. A 2200 pound vehicle maintains a constant 45 mph up a 3% incline. What horsepower is required?

(A) 5   (B) 8   (C) 24
(D) 72   (E) 114

---

10. The first derivative of kinetic energy with respect to velocity is

(A) force
(B) power
(C) kinetic energy squared
(D) momentum
(E) acceleration

---

11. A man weighs 210 pounds. He slides down a rope with a breaking strength of 160 pounds. What is the minimum downward acceleration if the rope is not to break?

(A) 2.4 ft/sec^2 (B) 3.9 ft/sec^2 (C) 7.7 ft/sec^2
(D) 15.8 ft/sec^2 (E) 27.3 ft/sec^2

---

12. The position of a particle which moves along a straight line is defined by the relation s = t^3 - 7t^2 - 13t + 45, where s is expressed in feet and t in seconds. At what time will the velocity be zero?

(A) 0 sec   (B) 32.76 sec   (C) 2.33 sec
(D) .79 sec   (E) 5.46 sec

---

13. A 45,000 pound railroad car is moving at a speed of 2 fps to the right. After colliding with a 40,000 pound car initially at rest, the 40,000 pound car moves to the right with a speed of 1.2 fps. What is the coefficient of restitution?

(A) 7.4   (B) .14   (C) 1.1
(D) .9   (E) .6

---

14. A motorist enters a curve of radius 400 feet with a speed of 60 mph. If the motorist increases the speed at a constant rate of 3 ft/sec^2, what will be the magnitude of the total acceleration after traveling 300 feet around the curve?

(A) 24 ft/sec^2 (B) 12 ft/sec^2 (C) 32 ft/sec^2
(D) 31 ft/sec^2 (E) 33 ft/sec^2

15. A projectile is fired with an initial velocity of 4000 fps as shown below. A vertical canyon wall rises 2000 feet as shown. The projectile impacts the edge of the cliff during the downward trajectory. What is the horizontal distance traveled by the projectile? (Assume constant gravity and the absence of air friction.)

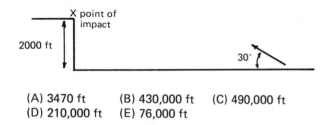

(A) 3470 ft   (B) 430,000 ft   (C) 490,000 ft
(D) 210,000 ft   (E) 76,000 ft

ENGINEER-IN-TRAINING
Mini-Exam

Mechanics of Materials

NOTICE TO EXAMINEE: This *MINI-EXAM* has been prepared to familiarize you with the types of questions that could appear on the Engineer-In-Training examination. If this *MINI-EXAM* is to be used as a measure of your speed and preparedness, it must <u>not</u> be taken prior to the completion of your review of the subject.

38

INSTRUCTIONS TO EXAMINEE: Please begin by entering your name, address, *MINI-EXAM* subject, and student number (if applicable) below.

All questions in this *MINI-EXAM* are multiple choice. After solving a problem, choose the closest answer given. Do not expect exact answers. Record your choice by blackening the appropriate circle below.

If it is your intent to take this *MINI-EXAM* under actual test conditions, you should allow yourself 30 minutes to complete all 15 questions. The time available per question under this time limitation is approximately the same as it will be in the actual Engineer-In-Training examination.

After you have completed the entire *MINI-EXAM*, turn to the solutions at the end of this booklet. Page, table, and figure numbers in the solutions refer to the *ENGINEER-IN-TRAINING REVIEW MANUAL*, 6th edition, by Michael R. Lindeburg.

MINI-EXAM subject: _____

1 (A) (B) (C) (D) (E)	6 (A) (B) (C) (D) (E)	11 (A) (B) (C) (D) (E)
2 (A) (B) (C) (D) (E)	7 (A) (B) (C) (D) (E)	12 (A) (B) (C) (D) (E)
3 (A) (B) (C) (D) (E)	8 (A) (B) (C) (D) (E)	13 (A) (B) (C) (D) (E)
4 (A) (B) (C) (D) (E)	9 (A) (B) (C) (D) (E)	14 (A) (B) (C) (D) (E)
5 (A) (B) (C) (D) (E)	10 (A) (B) (C) (D) (E)	15 (A) (B) (C) (D) (E)

Name_____ Student No. _____

Address_____

_____

1. $EI(d^3y/dx^3)$ can be solved without integration for

(A) shear
(B) moment
(C) deflection
(D) beam slope
(E) stress

2. If k is the radius of gyration, (L/k) is known as the

(A) moment of inertia
(B) statical moment
(C) slenderness ratio
(D) section modulus
(E) eccentricity ratio

3. The buckling force for a long column depends on

(A) the modulus of elasticity
(B) the cross sectional area
(C) the length
(D) the radius of gyration
(E) all of the above

4. A 2"x2" square beam experiences 1,030 ft-lbs of moment. What is the bending stress ½" from the beam's upper surface?

(A) 386 psi   (B) 773 psi   (C) 4640 psi
(D) 1545 psi   (E) 18,540 psi

5. A 20" square concrete piling is loaded with a 10,000 pound force along one centerline and x inches from the other centerline. What is the maximum value of x such that the concrete remains in compression everywhere?

(A) 33"   (B) 17"   (C) 3"
(D) 9"   (E) 14"

6. The deflection of a beam at some point x is y = ¼". What is the slope at that same point if the deflection curve is
$$y(x) = x^4/(1.14\ EE9)\ ?$$

(A) 1.0   (B) .0077   (C) .087
(D) 16,936   (E) 130

7. Aluminum used in a 1"x2" fixed-end column has a modulus of elasticity of 1 EE7 psi and a proportionality limit of 12,000 psi. What is the approximate minimum length for which Euler's column formula may be used?

(A) 25"   (B) 50"   (C) 150"
(D) 200"   (E) 300"

8. A 2 foot diameter cylindrical tank with .125" thick walls is constructed with spherical ends. If the stress is to be kept below 12,000 psi, what is the maximum pressure allowed in the tank?

(A) 250 psi   (B) 1500 psi   (C) 3000 psi
(D) 60 psi   (E) 125 psi

9. The force necessary to punch a ½" diameter hole through ½" plate steel (ultimate strength of 43,000 psi) is approximately

(A) 68 pounds   (B) 4220 pounds   (C) 34,000 pounds
(D) 56,000 pounds   (E) 68,000 pounds

10. A square steel bar 3"x3" is held rigidly between two walls as shown below. A force of 50,000 pounds loads the bar 3" from the left support. What is the approximate left-hand reaction if E = 3 EE7 psi?

(A) 10,000 lb   (B) 33,000 lb   (C) 25,000 lb
(D) 50,000 lb   (E) 15,000 lb

11. A steel bar (E = 3 EE7 psi) is stretched with a force of 50 pounds. If the bar is 4 feet long and has a cross-sectional area of 0.5 in^2, what is the approximate elongation?

(A) .0001 in   (B) .0002 in   (C) .0003 in
(D) .0004 in   (E) .0005 in

12. A 10 foot long vertical cable is fastened to the midpoint of a taut horizontal wire 12 feet long. When a weight is attached to the bottom of the vertical cable, the unit strains in the vertical cable and horizontal wire are .008 and .005 respectively. What is the approximate displacement of the weight?

(A) 4 in.   (B) 6 in.   (C) 7 in.
(D) 8 in.   (E) 9 in.

13. What is the maximum torque that can be applied to a 1 inch diameter shaft with a maximum stress of 8000 psi?

(A) 1200 in-lb   (B) 1500 in-lb   (C) 1570 in-lb
(D) 2000 in-lb   (E) 3100 in-lb

14. If the maximum permissible shearing stress is 10,000 psi, how many 3/8" bolts arranged in a 5" diameter bolt circle are required to transmit 100 hp at 315 rpm through a flanged coupling?

(A) 6   (B) 7   (C) 8
(D) 9   (E) 10

15. A round aluminum bar 1.000" in diameter and 10" long is to just fit into a steel collar when the bar is subjected to an axial compressive force of 10,000 lb. What should be the difference in bar outside diameter and collar inside diameter? (E = 1 EE7 psi, μ = .33)

(A) .0004"   (B) .0127"   (C) .0042"
(D) .00105"   (E) .00014"

ENGINEER-IN-TRAINING
Mini-Exam

DC Electricity

NOTICE TO EXAMINEE: This *MINI-EXAM* has been prepared to familiarize you with the types of questions that could appear on the Engineer-In-Training examination. If this *MINI-EXAM* is to be used as a measure of your speed and preparedness, it must <u>not</u> be taken prior to the completion of your review of the subject.

42

INSTRUCTIONS TO EXAMINEE: Please begin by entering your name, address, *MINI-EXAM* subject, and student number (if applicable) below.

All questions in this *MINI-EXAM* are multiple choice. After solving a problem, choose the closest answer given. Do not expect exact answers. Record your choice by blackening the appropriate circle below.

If it is your intent to take this *MINI-EXAM* under actual test conditions, you should allow yourself 30 minutes to complete all 15 questions. The time available per question under this time limitation is approximately the same as it will be in the actual Engineer-In-Training examination.

After you have completed the entire *MINI-EXAM*, turn to the solutions at the end of this booklet. Page, table, and figure numbers in the solutions refer to the *ENGINEER-IN-TRAINING REVIEW MANUAL*, 6th edition, by Michael R. Lindeburg.

MINI-EXAM subject: _____

1 Ⓐ Ⓑ Ⓒ Ⓓ Ⓔ	6 Ⓐ Ⓑ Ⓒ Ⓓ Ⓔ	11 Ⓐ Ⓑ Ⓒ Ⓓ Ⓔ
2 Ⓐ Ⓑ Ⓒ Ⓓ Ⓔ	7 Ⓐ Ⓑ Ⓒ Ⓓ Ⓔ	12 Ⓐ Ⓑ Ⓒ Ⓓ Ⓔ
3 Ⓐ Ⓑ Ⓒ Ⓓ Ⓔ	8 Ⓐ Ⓑ Ⓒ Ⓓ Ⓔ	13 Ⓐ Ⓑ Ⓒ Ⓓ Ⓔ
4 Ⓐ Ⓑ Ⓒ Ⓓ Ⓔ	9 Ⓐ Ⓑ Ⓒ Ⓓ Ⓔ	14 Ⓐ Ⓑ Ⓒ Ⓓ Ⓔ
5 Ⓐ Ⓑ Ⓒ Ⓓ Ⓔ	10 Ⓐ Ⓑ Ⓒ Ⓓ Ⓔ	15 Ⓐ Ⓑ Ⓒ Ⓓ Ⓔ

Name_____ Student No. _____

Address_____

_____

1. What is the current I in the curcuit shown below?

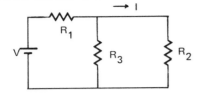

(A) $R_3/(R_1R_2+R_2R_3+R_1R_3)$
(B) $VR_3/(R_1R_2+R_1R_3+R_2R_3)$
(C) $VR_2/(R_1R_2+R_1R_3+R_2R_3)$
(D) $VR_2/(R_2+R_3)$
(E) $VR_3/(R_1+R_2+R_3)$

2. What is the current in the resistor R_3?

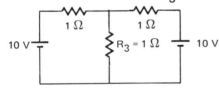

(A) 3.33 amps   (B) 6.67 amps   (C) 10.0 amps
(D) 13.33 amps   (E) 30.0 amps

3. What work is done in moving a charge of -x coulombs 2 meters in the same direction as a field of y volts/meter?

(A) -2xy Joules   (B) -2xy Newtons   (C) +2xy Joules
(D) + 2xy Newtons (E) -2xy volts/meter

4. What is I in the circuit shown below?

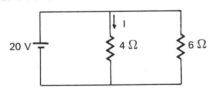

(A) 29 amps   (B) 36 amps   (C) 12 amps
(D) 5.0 amps   (E) 3.3 amps

5. A zener diode can be used in a circuit to provide a stable reference

(A) reluctance
(B) impedance
(C) resistance
(D) current
(E) voltage

6. A sphere of radius b with a charge of +q surrounds a sphere of radius a and charge -q. If r is a distance from the concentric center to a point between the two spheres, which of the following is proportional to the radial field?

(A) r   (B) r^2 (C) 1/r
(D) $1/(r^2)$ (E) $1/(r^3)$

7. A sphere of radius R has a uniform charge on its surface. The electric field at a distance r from the surface is proportional to

(A) r   (B) r^2 (C) $1/(r^2)$
(D) $(R+r)^2$ (E) $1/(R+r)^2$

8. The electrostatic flux between two concentric cylinders with radii a and b is proportional to

(A) r   (B) r^2 (C) 1/r
(D) $1/(r^2)$ (E) $1/(a+b)^2$

9. r is variable for the three-charge system shown below. At what distance r will the force on the middle charge be zero?

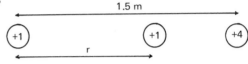

(A) .5 m   (B) 1.0 m   (C) .375 m
(D) .3 m   (E) .67 m

10. The input to a 3 dB amplifier is .001 watt. If the amplifier output feeds into a 2 dB attenuator, what will be the output power?

(A) .002 W   (B) .00125 W   (C) .0001 W
(D) .005 W   (E) .0008 W

11. A meter with a linear response reads 33% of full scale when 1.8 amps flow through it. If R_c is the coil resistance, what shunt resistance is required if the reading is to be 75% of full scale when 11 amps flow?

(A) $(24.7)R_c$ (B) $(1.7)R_c$ (C) $(.6)R_c$
(D) $(2.3)R_c$ (E) $(.44)R_c$

12. What is the magnitude of a constant current in a wire if 3000 coulombs of charge flow through the wire in 5 minutes?

(A) 600 amps   (B) 15,000 amps (C) 100 amps
(D) 10 amps   (E) .1 amps

13. What is the cost of delivering 3 horsepower to a motor for 24 hours if the motor is 95% efficient, requires 110 volts, and electricity costs $.10/kw-hr?

(A) $3.11   (B) $7.20   (C) $5.36
(D) $7.57   (E) $5.65

14. The resistance of a conductor is 20.3 ohms at 20°C and 25.9 ohms at 50°C. What is the approximate temperature coefficient of resistance? (1/°C)

(A) 9.20 EE-3   (B) .187   (C) 5.7 EE-2
(D) 3.7 EE-3   (E) 2.16 EE-3

15. A circular brass ring (permeability = 3.2 wb/amp-turn-m) has a circular cross section. The inside diameter of the ring is 4 inches and the outside diameter is 6 inches. If there are 2000 turns in the winding around the brass ring, what is the total flux produced by 2 amps flowing in the coil?

(A) 16 wb   (B) 43 wb   (C) 97 wb
(D) 140 wb   (E) 322 wb

ENGINEER-IN-TRAINING
Mini-Exam

AC Electricity

NOTICE TO EXAMINEE: This *MINI-EXAM* has been prepared to familiarize you with the types of questions that could appear on the Engineer-In-Training examination. If this *MINI-EXAM* is to be used as a measure of your speed and preparedness, it must <u>not</u> be taken prior to the completion of your review of the subject.

46

INSTRUCTIONS TO EXAMINEE: Please begin by entering your name, address, *MINI-EXAM* subject, and student number (if applicable) below.

All questions in this *MINI-EXAM* are multiple choice. After solving a problem, choose the closest answer given. Do not expect exact answers. Record your choice by blackening the appropriate circle below.

If it is your intent to take this *MINI-EXAM* under actual test conditions, you should allow yourself 30 minutes to complete all 15 questions. The time available per question under this time limitation is approximately the same as it will be in the actual Engineer-In-Training examination.

After you have completed the entire *MINI-EXAM*, turn to the solutions at the end of this booklet. Page, table, and figure numbers in the solutions refer to the *ENGINEER-IN-TRAINING REVIEW MANUAL*, 6th edition, by Michael R. Lindeburg.

MINI-EXAM subject: _____

1 (A) (B) (C) (D) (E)	6 (A) (B) (C) (D) (E)	11 (A) (B) (C) (D) (E)
2 (A) (B) (C) (D) (E)	7 (A) (B) (C) (D) (E)	12 (A) (B) (C) (D) (E)
3 (A) (B) (C) (D) (E)	8 (A) (B) (C) (D) (E)	13 (A) (B) (C) (D) (E)
4 (A) (B) (C) (D) (E)	9 (A) (B) (C) (D) (E)	14 (A) (B) (C) (D) (E)
5 (A) (B) (C) (D) (E)	10 (A) (B) (C) (D) (E)	15 (A) (B) (C) (D) (E)

Name_____ Student No. _____

Address_____

_____

1. Considering the circuit shown below, what is the turns ratio (n_1/n_2) for maximum power transfer?

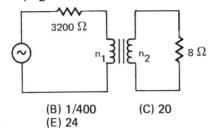

(A) 400   (B) 1/400   (C) 20
(D) 1/20   (E) 24

2. For the iron core transformer shown (which has $n_2 = 2n_1$), which of the following statements is true?

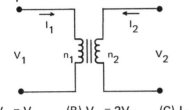

(A) $V_1 = V_2$ (B) $V_1 = 2V_2$ (C) $I_2 = 2I_1$
(D) $I_1 = 2I_2$ (E) $Z_2 = 2Z_1$

3. A sinusoidal voltage source has a peak value of 150 volts. What equivalent D.C. voltage source would produce the same heating effect in a 1 ohm resistor?

(A) 15 V   (B) 212 V   (C) 95 V
(D) 106 V   (E) 235 V

4. What is the admittance of the circuit shown below?

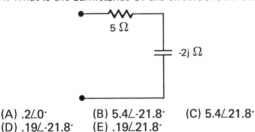

(A) $.2\angle 0°$ (B) $5.4\angle -21.8°$ (C) $5.4\angle 21.8°$
(D) $.19\angle -21.8°$ (E) $.19\angle 21.8°$

5. The switch closes at t=0. At what time t will be capacitor be 20% charged?

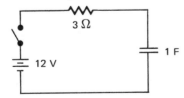

(A) 4.8 sec   (B) .67 sec   (C) .97 sec
(D) 6.67 sec   (E) 17.4 sec

6. What is the current in the inductor as a function of time after closing the switch?

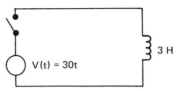

(A) $45t^2$ amps   (B) $10t^3$ amps   (C) 5 amps
(D) 10t amps   (E) $5t^2$ amps

7. What capacitance connected in series with a 50 ohm resistance will limit the current drawn from a 120-volt, 60 cycle source to 0.5 amps?

(A) 70.9 uF   (B) 11.3 uF   (C) 11.0 uF
(D) 4.2 mF   (E) 234.7 F

8. What is the impedance of the circuit shown below?

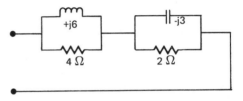

(A) $12\angle 76°$ (B) $4.3\angle 13°$ (C) $2.1\angle 11°$
(D) $9.6\angle 51°$ (E) $8.3\angle 21°$

9. A wye-connected, 3-phase, 440 volt alternator has a limit of 40 amps per coil. If the alternator supplies a line current of 25 amps at a power factor of .60, what is the power per phase?

(A) 6600 W   (B) 6350 W   (C) 5080 W
(D) 10,560 W   (E) 3810 W

10. The addition of gallium to pure germanium is an example of

(A) n-type doping
(B) biasing
(C) seeding
(D) p-type doping
(E) ionic bonding

11. A 120 volt generator delivers 30 KW to an electric furnace. The current supplied by the generator is about

(A) 500 amps   (B) 4 amps   (C) 100 amps
(D) 250 amps   (E) 360 amps

12. A 10 ohm resistor is connected in series with a capacitive reactance of 24 ohms. What is the total impedance?

(A) $26[e^{-j1.18}]$ (B) $13[\cos(5) - j\sin(12)]$
(C) $[5 + j12]$ (D) $26\angle -43.2°$
(E) $13[e^{j1.18}]$

48

13. Initially, C_1 has a charge of 600 coulombs and C_2
is uncharged. What is the current in the resistor at the
instant the switch is closed?

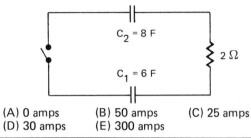

(A) 0 amps   (B) 50 amps   (C) 25 amps
(D) 30 amps   (E) 300 amps

14. An industrial plant operates with an average load
of 500 KW. The power factor averages 70% lagging.
A 150 KW synchronous motor is added to the plant
load. What is the power factor at which the motor
must operate to raise the power factor to 85%?

(A) .687   (B) .715   (C) .810
(D) .950   (E) .876

15. What is the apparent power of a pure 1.0 H
inductance drawing 2 amps from a 60 cycle source?

(A) 240 VA   (B) 754 VA   (C) 1508 VA
(D) 915 VA   (E) 1973 VA

ENGINEER-IN-TRAINING
Mini-Exam

Materials Science

NOTICE TO EXAMINEE: This *MINI-EXAM* has been prepared to familiarize you with the types of questions that could appear on the Engineer-In-Training examination. If this *MINI-EXAM* is to be used as a measure of your speed and preparedness, it must <u>not</u> be taken prior to the completion of your review of the subject.

50

INSTRUCTIONS TO EXAMINEE: Please begin by entering your name, address, *MINI-EXAM* subject, and student number (if applicable) below.

All questions in this *MINI-EXAM* are multiple choice. After solving a problem, choose the closest answer given. Do not expect exact answers. Record your choice by blackening the appropriate circle below.

If it is your intent to take this *MINI-EXAM* under actual test conditions, you should allow yourself 30 minutes to complete all 15 questions. The time available per question under this time limitation is approximately the same as it will be in the actual Engineer-In-Training examination.

After you have completed the entire *MINI-EXAM*, turn to the solutions at the end of this booklet. Page, table, and figure numbers in the solutions refer to the *ENGINEER-IN-TRAINING REVIEW MANUAL*, 6th edition, by Michael R. Lindeburg.

MINI-EXAM subject: _____

1 (A) (B) (C) (D) (E)   6 (A) (B) (C) (D) (E)   11 (A) (B) (C) (D) (E)
2 (A) (B) (C) (D) (E)   7 (A) (B) (C) (D) (E)   12 (A) (B) (C) (D) (E)
3 (A) (B) (C) (D) (E)   8 (A) (B) (C) (D) (E)   13 (A) (B) (C) (D) (E)
4 (A) (B) (C) (D) (E)   9 (A) (B) (C) (D) (E)   14 (A) (B) (C) (D) (E)
5 (A) (B) (C) (D) (E)   10 (A) (B) (C) (D) (E)   15 (A) (B) (C) (D) (E)

Name_____ Student No. _____

Address_____

_____

1. What is the toughness of the material whose stress-strain curve is shown below?

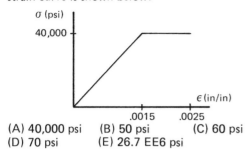

(A) 40,000 psi   (B) 50 psi   (C) 60 psi
(D) 70 psi   (E) 26.7 EE6 psi

2. Impact tests are used to determine a material's

(A) elasticity
(B) Poisson's ratio
(C) endurance limit
(D) hardness
(E) transition temperature

3. Given the binary phase diagram shown below, what is the concentration of alloy B at the eutectoid point?

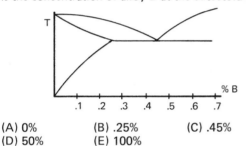

(A) 0%   (B) .25%   (C) .45%
(D) 50%   (E) 100%

4. How many atoms are contained in one unit cell of a body-centered cubic lattice?

(A) 1   (B) 2   (C) 3
(D) 5   (E) 9

5. At temperature T^*, what phases are present in an alloy with a% of the alloying element?

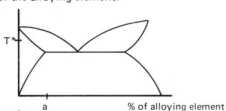

(A) alpha only
(B) liquid only
(C) liquid and alpha solid
(D) liquid and beta solid
(E) alpha solid and beta solid

6. Which of the following is used to determine inter-atomic spacing in a crystalline material?

(A) allotropic changes
(B) austenitizing
(C) x-ray diffraction
(D) neutron diffraction
(E) spectroscopy

7. Assume the density of aluminum is 2.7 g/cc. Approximately how many atoms are present in one cubic centimeter of aluminum?

(A) 2.9 EE23   (B) 7.83 EE23   (C) 6.02 EE22
(D) 1.63 EE24   (E) 6.02 EE23

8. Which of the following is not a body-centered cubic crystal?

(A) chromium   (B) cadmium   (C) lithium
(D) sodium   (E) tantalum

9. Which of the following describes a material that possesses different properties when stressed in different directions?

(A) atactic
(B) anisotropic
(C) isotropic
(D) inclusive
(E) asymmetrical

10. What are the majority carriers in an n-type semiconductor?

(A) neutrons
(B) electrons
(C) holes
(D) protons
(E) positrons

11. 20,000 psi is often taken as an allowable stress for structural steel. What does this limit imply?

(A) The steel will yield if subjected to a tensile stress greater than 20,000 psi.
(B) The ultimate strength of the steel is 20,000 psi.
(C) 20,000 psi is the yield strength.
(D) Alloy steels cannot be safely stressed higher than 20,000 psi.
(E) None of the above.

12. The hardness of steel may be increased by heating to approximately 1500·F and quenching in oil or water if

(A) the carbon content is between 0.2% and 2.0%
(B) the carbon content is above 3%
(C) the carbon content is less than 0.2%
(D) all carbon is removed and the steel only contains chromium, nickel, or manganese
(E) the steel has been rolled, not cast

13. A steel rod is stressed in tension slightly beyond its yield point so that it suffers permanent deformation without rupture. If the rod is released and then again stressed in tension,

(A) the yield point will be the same as for the first tensioning.
(B) the ductility of the steel will be higher.
(C) the ultimate strength will be lower.
(D) the rod will break at a stress which is less than the original yield strength.
(E) a new yield point will be established which is higher than the original yield point.

52

14. A contractor adds extra water to his concrete to make a more workable mix. Which of the following statements is true?

(A) Excess water has little effect on the strength of concrete.
(B) As long as the aggregates are not allowed to settle, the concrete will not be affected by excess water.
(C) Once the concrete is poured in the forms, it can be vibrated to mix up the aggregates and the water that comes to the surface can be drained off with little effect on the concrete strength.
(D) Excess water in concrete lowers the strength of concrete even after the pour has completely cured.
(E) The only reason that excess water is detrimental is because it causes the wooden forms to warp, and much concrete is lost through the resulting cracks.

15. The endurance limit of steel is

(A) equal to the yield strength.
(B) equal to the ultimate tensile strength.
(C) equal to approximately half of the ultimate tensile strength.
(D) equal to approximately half of the elastic limit.
(E) none of the above.

ENGINEER-IN-TRAINING
Mini-Exam

Peripheral Sciences

NOTICE TO EXAMINEE: This *MINI-EXAM* has been prepared to familiarize you with the types of questions that could appear on the Engineer-In-Training examination. If this *MINI-EXAM* is to be used as a measure of your speed and preparedness, it must <u>not</u> be taken prior to the completion of your review of the subject.

54

INSTRUCTIONS TO EXAMINEE: Please begin by entering your name, address, *MINI-EXAM* subject, and student number (if applicable) below.

All questions in this *MINI-EXAM* are multiple choice. After solving a problem, choose the closest answer given. Do not expect exact answers. Record your choice by blackening the appropriate circle below.

If it is your intent to take this *MINI-EXAM* under actual test conditions, you should allow yourself 30 minutes to complete all 15 questions. The time available per question under this time limitation is approximately the same as it will be in the actual Engineer-In-Training examination.

After you have completed the entire *MINI-EXAM*, turn to the solutions at the end of this booklet. Page, table, and figure numbers in the solutions refer to the *ENGINEER-IN-TRAINING REVIEW MANUAL*, 6th edition, by Michael R. Lindeburg.

MINI-EXAM subject: _____

1 (A) (B) (C) (D) (E)   6 (A) (B) (C) (D) (E)   11 (A) (B) (C) (D) (E)
2 (A) (B) (C) (D) (E)   7 (A) (B) (C) (D) (E)   12 (A) (B) (C) (D) (E)
3 (A) (B) (C) (D) (E)   8 (A) (B) (C) (D) (E)   13 (A) (B) (C) (D) (E)
4 (A) (B) (C) (D) (E)   9 (A) (B) (C) (D) (E)   14 (A) (B) (C) (D) (E)
5 (A) (B) (C) (D) (E)   10 (A) (B) (C) (D) (E)   15 (A) (B) (C) (D) (E)

Name_____ Student No. _____

Address_____

_____

1. Heat is conducted at 1000 BTU/hr through a 1 foot thick rectangular 2' x 3' plane. What is the thermal conductivity if the temperature gradient is 460·F? (All answers are in BTU/hr-ft-·F.)

(A) 6.2   (B) 1.37   (C) .36
(D) 3.61   (E) .01

2. Which of the following is the easiest method of determining the relative humidity of an air sample?

(A) Measure the wet and dry bulb temperatures and refer to a psychrometric chart.
(B) Measure the wet and dry bulb temperatures and divide the latter by the former.
(C) Measure the wet and dry bulb temperatures and refer to a Mollier diagram.
(D) Measure the barometric pressure and the dry bulb temperature and divide the former by the latter.
(E) Measure the barometric pressure and refer to a psychrometric chart.

3. Everything else remaining constant, the heat radiated from a surface

(A) varies as the 4th power of the surface's absolute temperature.
(B) varies inversely as the temperature of the surface.
(C) varies directly with the absolute temperature of the surface.
(D) varies directly with the square of the absolute temperature of the surface.
(E) none of the above

4. Which of the following is a definition of relative humidity?

(A) The weight of moisture per cubic foot of dry air.
(B) The ratio of vapor pressure to the saturation pressure for that temperature.
(C) The wet bulb temperature divided by the dry bulb temperature.
(D) The specific heat of the moisture divided by the specific heat of the air.
(E) none of the above

5. A finned heat exchanger surrounded by air carries high temperature steam. Why should the fins be placed on the outside instead of the inside?

(A) High temperature steam is too corrosive for thin metal fins.
(B) The fins serve as a dust filter.
(C) The fins increase the heat transfer coefficient of the air.
(D) The primary resistance to heat flow is on the outside, and the fins reduce this resistance.
(E) Placement of the fins is not critical. However, external fins are easier to manufacture.

6. A 9" x 9" contact print aerial photograph is taken at a height of 10,000 feet with a camera whose focal length is 6". What geographical area is shown?

(A) 4000' x 4000'   (B) 9000' x 9000'
(C) 3 miles x 3 miles   (D) 6 miles x 6 miles
(E) 8 miles x 8 miles

7. Which of the following is the best thermal insulator?

(A) glass   (B) wood (fir)   (C) concrete
(D) soil   (E) steel

8. The rate of heat transfer through a given section of a homogeneous wall is

(A) directly proportional to the coefficient of heat transfer and the wall thickness.
(B) inversely proportional to the coefficient of heat transfer and directly proportional to the wall thickness.
(C) directly proportional to the coefficient of heat transfer and inversely proportional to the wall thickness.
(D) inversely proportional to the coefficient of heat transfer and to the wall thickness.
(E) independent of the wall thickness.

9. A one candlepower light source radiates luminous flux at the rate of

(A) 1 lumen   (B) 2 lumens   (C) π lumens
(D) 2π lumens   (E) 4π lumens

10. What is the frequency of a light ray which has a wavelength of 2 EE-5 inch?

(A) 300,000 GHz   (B) 600,000 GHz   (C) 1,180,000 GHz
(D) 1,200,000 GHz   (E) 6,000,000 GHz

11. One angstrom unit is

(A) EE-2 meters   (B) EE-4 meters   (C) EE-6 meters
(D) EE-8 meters   (E) EE-10 meters

12. When air with a 50% relative humidity is heated from 46·F to 75·F at constant moisture,

(A) the relative humidity increases.
(B) the dew point is lowered.
(C) the heat absorbed is equal to the increase in the air mixture's enthalpy.
(D) the heat absorbed is equal to the increase in heat content of the air only since the moisture heat is constant.
(E) none of the above

13. Water hardness is primarily due to

(A) carbonates and sulphates of calcium and magnesium
(B) alum
(C) soda ash
(D) sodium sulphate
(E) sodium chloride

56

14. The concentration or strength of waste water is measured by its

(A) hydrogen content
(B) biochemical oxygen demand
(C) nitrogen-protein ratio
(D) hydrogen ion concentration
(E) titration ratio

15. Which of the following is not directly or indirectly measured by an Orsat analysis?

(A) CO (B) CO_2 (C) O_2
(D) N_2 (E) H_2S

ENGINEER-IN-TRAINING
Mini-Exam

Systems Theory

NOTICE TO EXAMINEE: This *MINI-EXAM* has been prepared to familiarize you with the types of questions that could appear on the Engineer-In-Training examination. If this *MINI-EXAM* is to be used as a measure of your speed and preparedness, it must <u>not</u> be taken prior to the completion of your review of the subject.

58

INSTRUCTIONS TO EXAMINEE: Please begin by entering your name, address, *MINI-EXAM* subject, and student number (if applicable) below.

All questions in this *MINI-EXAM* are multiple choice. After solving a problem, choose the closest answer given. Do not expect exact answers. Record your choice by blackening the appropriate circle below.

If it is your intent to take this *MINI-EXAM* under actual test conditions, you should allow yourself 30 minutes to complete all 15 questions. The time available per question under this time limitation is approximately the same as it will be in the actual Engineer-In-Training examination.

After you have completed the entire *MINI-EXAM*, turn to the solutions at the end of this booklet. Page, table, and figure numbers in the solutions refer to the *ENGINEER-IN-TRAINING REVIEW MANUAL*, 6th edition, by Michael R. Lindeburg.

MINI-EXAM subject: _____

1 (A) (B) (C) (D) (E)   6 (A) (B) (C) (D) (E)   11 (A) (B) (C) (D) (E)
2 (A) (B) (C) (D) (E)   7 (A) (B) (C) (D) (E)   12 (A) (B) (C) (D) (E)
3 (A) (B) (C) (D) (E)   8 (A) (B) (C) (D) (E)   13 (A) (B) (C) (D) (E)
4 (A) (B) (C) (D) (E)   9 (A) (B) (C) (D) (E)   14 (A) (B) (C) (D) (E)
5 (A) (B) (C) (D) (E)   10 (A) (B) (C) (D) (E)   15 (A) (B) (C) (D) (E)

Name_____ Student No. _____

Address_____

_____

1. What is the Laplace transform of a step function whose height is h?

(A) h   (B) s   (C) h/s
(D) s/h   (E) 1/(s+h)

2. What is the overall system gain through the feedback system shown below?

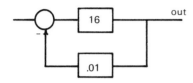

(A) 1.19   (B) 1600   (C) .16
(D) 13.79   (E) 19.05

3. What is the initial value of the function f(t) which corresponds to

$$F(s) = \frac{2(s-1)}{s^2-2s+2}$$

(A) 0   (B) +1   (C) +2
(D) +∞   (E) -∞

4. Which of the values below comes closest to a pole of

$$F(s) = \frac{2(s+4)}{s(s-3)}$$

(A) -4   (B) 2   (C) 3
(D) 6   (E) 4/3

5. What is the final value of f(t) corresponding to

$$F(s) = \frac{(s+4)(s^2-2s+2)}{(s+4)^2}$$

(A) 0   (B) ½   (C) 16
(D) +∞   (C) -∞

6. Which of the values below comes closest to a zero of

$$F(s) = \frac{2(s+4)}{s(s-3)}$$

(A) +4   (B) +3   (C) -4
(D) -3   (E) 8/3

7. The f(t) corresponding to the F(s) given below is

$$F(s) = s/(s^2+4)$$

(A) ½sin(2t)   (B) e^{-4t} (C) 2tan(4t)
(D) 0   (E) cos(2t)

8. What is the gain of the cascaded system shown?

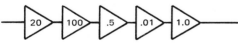

(A) 20   (B) .1   (C) .0001
(D) 100   (E) 10

9. A linear programming problem with only two independent variables

(A) cannot be solved without a computer.
(B) should be solved by trial and error.
(C) can be solved graphically.
(D) can be solved by inspection.
(E) must be solved iteratively.

10. Positive feedback is

(A) more desireable than negative feedback.
(B) more stable than negative feedback.
(C) results in an output which is less sensitive to changes in the input.
(D) may be used with electrical systems but not with mechanical systems.
(E) none of the above.

11. What is the gain or transfer function of the system shown?

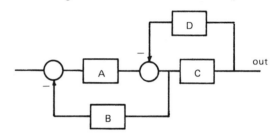

(A) $\frac{AC}{1+DC+BA}$ (B) $\frac{AC}{1-DC+BA}$ (C) $\frac{AC}{1-DC-BA}$

(D) AC/(1+DB)   (E) AC/(1-DB)

12. The mechanical response to a repeating sinusoidal input force will probably be

(A) sinusoidal
(B) a square wave
(C) a decaying sinusoid
(D) exponential
(E) an impulse

13. Stability of systems can be evaluated by use of

(A) the Bernoulli equation.
(B) the Lagrangian polynomial.
(C) the method of golden sections.
(D) iterative approximations
(E) the Routh criterion.

14. For a system to be stable, the root locus plot must not have any poles

(A) in the top half of the plane.
(B) in the bottom half of the plane.
(C) in the left half of the plane.
(D) in the right half of the plane.
(E) at all.

15. The waveform shown below can be easily evaluated by means of Fourier analysis because it has

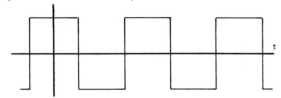

(A) even function symmetry.
(B) odd function symmetry.
(C) rotational symmetry.
(D) quarter-wave symmetry.
(E) square symmetry.

ENGINEER-IN-TRAINING Mini-Exam

Computer Science

NOTICE TO EXAMINEE: This *MINI-EXAM* has been prepared to familiarize you with the types of questions that could appear on the Engineer-In-Training examination. If this *MINI-EXAM* is to be used as a measure of your speed and preparedness, it must <u>not</u> be taken prior to the completion of your review of the subject.

62

INSTRUCTIONS TO EXAMINEE: Please begin by entering your name, address, *MINI-EXAM* subject, and student number (if applicable) below.

All questions in this *MINI-EXAM* are multiple choice. After solving a problem, choose the closest answer given. Do not expect exact answers. Record your choice by blackening the appropriate circle below.

If it is your intent to take this *MINI-EXAM* under actual test conditions, you should allow yourself 30 minutes to complete all 15 questions. The time available per question under this time limitation is approximately the same as it will be in the actual Engineer-In-Training examination.

After you have completed the entire *MINI-EXAM*, turn to the solutions at the end of this booklet. Page, table, and figure numbers in the solutions refer to the *ENGINEER-IN-TRAINING REVIEW MANUAL*, 6th edition, by Michael R. Lindeburg.

MINI-EXAM subject: _____

1 Ⓐ Ⓑ Ⓒ Ⓓ Ⓔ	6 Ⓐ Ⓑ Ⓒ Ⓓ Ⓔ	11 Ⓐ Ⓑ Ⓒ Ⓓ Ⓔ
2 Ⓐ Ⓑ Ⓒ Ⓓ Ⓔ	7 Ⓐ Ⓑ Ⓒ Ⓓ Ⓔ	12 Ⓐ Ⓑ Ⓒ Ⓓ Ⓔ
3 Ⓐ Ⓑ Ⓒ Ⓓ Ⓔ	8 Ⓐ Ⓑ Ⓒ Ⓓ Ⓔ	13 Ⓐ Ⓑ Ⓒ Ⓓ Ⓔ
4 Ⓐ Ⓑ Ⓒ Ⓓ Ⓔ	9 Ⓐ Ⓑ Ⓒ Ⓓ Ⓔ	14 Ⓐ Ⓑ Ⓒ Ⓓ Ⓔ
5 Ⓐ Ⓑ Ⓒ Ⓓ Ⓔ	10 Ⓐ Ⓑ Ⓒ Ⓓ Ⓔ	15 Ⓐ Ⓑ Ⓒ Ⓓ Ⓔ

Name_____ Student No. _____

Address_____

_____

1. Which of the following is not a component of a computer?

(A) control unit
(B) I/O devices
(C) memory
(D) logical unit
(E) power supply

2. Convert $(1101111101)_2$ to base 16.

(A) 1575   (B) 67D   (C) 747
(D) DF1   (E) 37D

3. A bootstrap loader is

(A) a short sequence of instructions which loads a larger program.
(B) a form of compiler.
(C) a read-only memory containing programs for input/output devices.
(D) a program which loads application programs.
(E) a program which controls the paper tape reader.

4. How many bits are contained in a 12-byte instruction?

(A) 48   (B) 144   (C) ¾
(D) 72   (E) 96

5. The code used with punched cards is

(A) ASCII
(B) binary coded decimal (BCD)
(C) Hollerith
(D) EBCDIC
(E) hexadecimal

6. Which of the following is not a higher-level language?

(A) assembly language
(B) FORTRAN
(C) COBOL
(D) PL/I
(E) BASIC

7. What is the value of the FORTRAN variable WYE at the end of the sequence given below?

```
INTEGER WYE*2
STAR = 14.7
WYE = (STAR/4)*3/2
```

(A) 4   (B) 5.51   (C) 7.6
(D) 6   (E) 5

8. What is the value of $(75)_{10}$ in base 4?

(A) 3201   (B) 1101101   (C) 2313
(D) 1344   (E) 1023

9. What is the output voltage if the input is 2 volts?

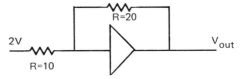

(A) -2   (B) $\approx$-EE5   (C) -4
(D) -1   (E) -400

10. What does the assembly language instruction given below do?

```
LA  1,X
```

(A) move the decimal point of variable X one point to the left
(B) read the last value of X and put it into register 1
(C) establish register 1 as the storage location for logical variable X
(D) multiply the number in register 1 by variable LA
(E) load register 1 with the variable X

11. What type of FORTRAN variable would appear in a program as 'S(5)'?

(A) integer dimension
(B) fixed-point integer
(C) logical variable
(D) conditional counter
(E) floating-point array member

12. How would variable X be printed out as a result of the following sequence of FORTRAN statements?

```
X = 59432.79
J = FIX(X)
WRITE J(5,10)
10 FORMAT (I4)
```

(A) 9432   (B) 5943   (C) 59433
(D) 59432   (E) ****

13. The BASIC language

(A) exists in many forms.
(B) is used in industrial systems microprocessors.
(C) is ideal for string manipulation.
(D) is similar to assembly language.
(E) cannot be used for mathematical calculations other than addition, subtraction, multiplication, and division.

14. What operation is performed by the operational amplifier shown below?

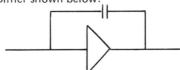

(A) summation   (B) scalar multiplication   (C) inversion
(D) integration   (E) differentiation

15. What is the value of X in the FORTRAN statement below?

$$X = (5.7)*(2.1)/(9.9)+(1.3)**(6.1)$$

(A) .806 (B) 6.16 (C) 1.50
(D) 9.20 (E) 3.85

ENGINEER-IN-TRAINING
Mini-Exam

Nucleonics

INSTRUCTIONS TO EXAMINEE: Please begin by entering your name, address, *MINI-EXAM* subject, and student number (if applicable) below.

All questions in this *MINI-EXAM* are multiple choice. After solving a problem, choose the closest answer given. Do not expect exact answers. Record your choice by blackening the appropriate circle below.

If it is your intent to take this *MINI-EXAM* under actual test conditions, you should allow yourself 30 minutes to complete all 15 questions. The time available per question under this time limitation is approximately the same as it will be in the actual Engineer-In-Training examination.

After you have completed the entire *MINI-EXAM*, turn to the solutions at the end of this booklet. Page, table, and figure numbers in the solutions refer to the *ENGINEER-IN-TRAINING REVIEW MANUAL*, 6th edition, by Michael R. Lindeburg.

MINI-EXAM subject: _____

```
 1  Ⓐ Ⓑ Ⓒ Ⓓ Ⓔ      6  Ⓐ Ⓑ Ⓒ Ⓓ Ⓔ     11  Ⓐ Ⓑ Ⓒ Ⓓ Ⓔ
 2  Ⓐ Ⓑ Ⓒ Ⓓ Ⓔ      7  Ⓐ Ⓑ Ⓒ Ⓓ Ⓔ     12  Ⓐ Ⓑ Ⓒ Ⓓ Ⓔ
 3  Ⓐ Ⓑ Ⓒ Ⓓ Ⓔ      8  Ⓐ Ⓑ Ⓒ Ⓓ Ⓔ     13  Ⓐ Ⓑ Ⓒ Ⓓ Ⓔ
 4  Ⓐ Ⓑ Ⓒ Ⓓ Ⓔ      9  Ⓐ Ⓑ Ⓒ Ⓓ Ⓔ     14  Ⓐ Ⓑ Ⓒ Ⓓ Ⓔ
 5  Ⓐ Ⓑ Ⓒ Ⓓ Ⓔ     10  Ⓐ Ⓑ Ⓒ Ⓓ Ⓔ     15  Ⓐ Ⓑ Ⓒ Ⓓ Ⓔ
```

Name_____ Student No. _____

Address_____

_____

1. One fission yields 180 EE6 electron volts. If one electron volt equals 1.6 EE-19 joules, what fission rate is necessary to generate one watt?

(A) 2.9 EE-11 fissions per second
(B) 3.5 EE10 fissions per second
(C) 4.5 EE18 fissions per second
(D) 6.3 EE18 fissions per second
(E) 1.1 EE25 fissions per second

2. U_{235} and U_{238} have the same number of

(A) neutrons
(B) electrons
(C) photons
(D) nucleons
(E) isotopes

3. Deuterium differs from hydrogen in the number of

(A) neutrons
(B) electrons
(C) protons
(D) isotopes
(E) mass defects

4. The density of $_{26}Fe^{56}$ is 7.86 g/cc. What is the approximate volume of an iron atom?

(A) .8 EE-23 cubic centimeter
(B) 1.2 EE-23 cubic centimeter
(C) 7.3 EE-23 cubic centimeter
(D) 9.3 EE-23 cubic centimeter
(E) 43.0 EE-23 cubic centimeter

5. How many half-lives are required to reduce the remaining part of a radioactive isotope to less than 1% of the original?

(A) 7   (B) 8   (C) 6
(D) 5   (E) 9

6. Which of the following occurs in the nuclear reaction given below?

$$_7N^{13} \rightarrow {_6}C^{13}$$

(A) A beta ray is emitted.
(B) An alpha particle is emitted.
(C) A positron is emitted.
(D) A neutron is emitted.
(E) An x-ray is captured.

7. The radius of an electron orbit is known to be .80 angstroms. What is the de Broglie wavelength of the electron if 6 complete cycles constitute a stable pattern around the nucleus?

(A) .91 Å   (B) 9.60 Å   (C) 5.02 Å
(D) .13 Å   (E) .84 Å

8. What is the energy of an electron traveling at .85 the speed of light?

(A) 6.53 EE-10 joules
(B) 1.56 EE-13 joules
(C) 23 EE-14 joules
(D) 9.38 EE13 joules
(E) 9.64 EE-14 joules

9. A photon is

(A) a neutral proton
(B) a gamma ray
(C) a light particle
(D) a charged neutron
(E) electromagnetic energy

10. An x-ray is

(A) a high-speed electron
(B) a high-speed neutron
(C) a high-speed proton
(D) a high-speed positron
(E) electromagnetic radiation

11. Which of the following are used as fuels in nuclear reactors?

(A) uranium only
(B) plutonium only
(C) sodium only
(D) uranium and plutonium
(E) uranium, plutonium, and sodium

12. The initial activity of a radioactive substance is 1 curie. The half-life of the substance is 15 hours. How long will it take before the activity is .25 curies?

(A) 21.6 hrs   (B) 3.75 hrs   (C) 60 hrs
(D) 30 hrs   (E) 40 hrs

13. The group of metals which contains lithium, sodium, potassium, rubidium, and cesium forms a closely-related family known as

(A) the rare earth group.
(B) the metals of the fourth outer group.
(C) the alkalai metals
(D) the elements of the inner group
(E) the metals of group VIII

14. A mu meson (muon) has a mass approximately

(A) equal to that of an electron
(B) equal to that of a proton
(C) equal to that of a neutron
(D) between the masses of an electron and a proton
(E) between the masses of a proton and an alpha particle

15. Find the age of a wood beam taken from an ancient ruin if 40% of the original carbon-14 is still present. The half-life of carbon-14 is approximately 5700 years.

(A) 8225 years   (B) 14,250 years   (C) 4560 years
(D) 7535 years   (E) 7980 years

MATHEMATICS

1 EXPAND BY THE FIRST COLUMN

$$6\begin{vmatrix} 6 & 1 \\ -1 & -2 \end{vmatrix} - 9\begin{vmatrix} -4 & 2 \\ -1 & -2 \end{vmatrix} + 3\begin{vmatrix} -4 & 2 \\ 6 & 1 \end{vmatrix}$$

$$= 6(-12+1) - 9(8+2) + 3(-4-12)$$
$$= 6(-11) - 9(10) + 3(-16) = -66 - 90 - 48$$
$$= -204$$

$\boxed{C}$

2 FIND THE ROOTS OF THE CHARACTERISTIC EQUATION

$$R^2 + 4R + 4 = 0$$
$$(R+2)(R+2) = 0$$
$$R_1 = -2, \quad R_2 = -2$$
$$y = A_1 e^{-2t} + A_2 t e^{-2t}$$

$\boxed{B}$

3
$$\int_2^5 \frac{1}{x^2} dx = \int_2^5 (x^{-2}) dx = -x^{-1} \Big]_2^5$$
$$= -\frac{1}{5} + \frac{1}{2} = .3$$

$\boxed{D}$

4 SINCE NEITHER THE NUMERATOR NOR DENOMINATOR GO TO ZERO OR INFINITY, THE 4 CAN BE SUBSTITUTED DIRECTLY FOR X

$$\frac{4((4)^3 - 6)}{4-5} = \frac{4(64-6)}{-1} = -232$$

$\boxed{D}$

5 THIS IS NOT A CUBIC, SO IT CANNOT BE A 3-D SPHERE. BECAUSE OF THE -4/X AND +2Y TERMS IT CANNOT BE CENTERED AT THE ORIGIN. FACTORING,

$$(x-2)^2 + (y+1)^2 = 20+4+1 = 25$$

THIS IS A CIRCLE OF RADIUS 5 CENTERED AT $(2, -1)$

$\boxed{D}$

6 THE SLOPE OF THE PERPENDICULAR LINE MUST BE $(-1/3)$. THE INTERCEPTS ARE NOT RELEVANT.

$$9y = -3x + 17$$
$$y = -\frac{1}{3}x + \frac{17}{9}$$

$\boxed{D}$

7
$$z = \frac{1}{\sqrt{2k^2+8}} = (2k^2+8)^{-1/2}$$
$$\frac{dz}{dk} = \left(-\frac{1}{2}\right)(2k^2+8)^{-3/2}(4k) = \frac{-2k}{\sqrt{(2k^2+8)^3}}$$

$\boxed{B}$

8
$$f'(x) = x^2 + x - 12 = 0$$
THIS HAS ROOTS OF -4 AND +3.
$$f''(x) = 2x + 1$$
AT X=3, f'' IS POSITIVE $\longrightarrow$ MINIMUM
$$f(3) = \frac{1}{3}(3)^3 + \frac{1}{2}(3)^2 - 12(3) + 34$$
$$= 11.5$$

$\boxed{E}$

9
$$\left(\frac{3}{10}\right)\left(\frac{6}{10}\right) + \left(\frac{4}{10}\right)\left(\frac{7}{10}\right)$$
$$= \frac{18+28}{100} = \frac{46}{100} = \frac{23}{50}$$

$\boxed{E}$

10 SOLVE FOR THE ROOTS OF THE CHARACTERISTIC EQUATION.

$$R^2 + 4R - 12 = 0$$
$$(R+6)(R-2) = 0$$
$$x = A_1 e^{-6t} + A_2 e^{2t}$$

ALTHOUGH CHOICE C WOULD WORK IF $A_1 = 0$, WE CHOOSE

$\boxed{B}$

11 DIVIDE BOTH NUMERATOR AND DENOMINATOR BY X. $\quad \frac{1}{4}x - 1 + \frac{3}{4x}$

AS $X \rightarrow 0$, THIS APPROACHES ∞

$\boxed{B}$

12 ALTHOUGH COMBINATIONS COULD BE USED, THE EASIEST WAY IS TO COMPUTE

$$2^5 - 1 = 31$$

$\boxed{A}$

13
$$\int_0^2 3x^2 dx = x^3 \Big]_0^2 = 8$$

$\boxed{E}$

14 SINCE THE 1ST DERIVATIVE WILL HAVE ONE OR MORE TERMS WITH $(1+x^2)$ IN THE DENOMINATOR, THE 1ST DERIVATIVE CANNOT EQUAL ZERO IN THE INTERVAL.

$$f(x) = x^2(1+x^2)^{-1}$$
$$f'(x) = x^2(-1)(1+x^2)^{-2}(2x) + (1+x^2)^{-1}(2x)$$
$$= \frac{-2x^3}{(1+x^2)^2} + \frac{2x}{(1+x^2)}$$

THE ONLY WAY $f'(x)$ CAN BE ZERO IS FOR X TO BE ZERO, WHICH IS NOT IN THE INTERVAL.

CHECK THE ENDPOINTS.

$$f(9) = \frac{81}{82} \qquad f(10) = \frac{100}{101}$$

9 IS THE MINIMUM

$\boxed{C}$

15 ASSUME POISSON DISTRIBUTION

$$P\{8\} = \frac{e^{-20} \, 20^8}{8!} = .001309$$

$\boxed{A}$

ENGINEERING ECONOMIC ANALYSIS

1 $750,000 - \frac{1}{2}(750,000) - 500,000$

$= -125,000$ LOSS

$\boxed{A}$

2 $\frac{13}{2} = 6.5\%$

6 YEARS = 12 SEMI-ANNUAL COMPOUNDING PERIODS

$(27,000)(1 + .065)^{-12} = 12681$

$\boxed{C}$

3 $T = \frac{1}{2}(10)(11) = 55$

$D = \frac{6}{55}(6600 - 1600) = 545.45$

$\boxed{E}$

4 $\left(1 + \frac{.075}{4}\right)^4 - 1 = .077135$

$\boxed{C}$

5 $P = 5600(P/A, 10\%, 10)$

$= 5600(6.1446) = 34409.76$

$\boxed{E}$

6 THE EUAC OF THE FIRST CYCLE IS

$10,000(A/P, 8\%, 10) + 100 + 500(P/F, 8\%, 5)(A/P, 8\%, 10)$

$= 10,000(.1490) + 100 + 500(.6806)(.1490)$

$= 1640.70$

CAPITALIZED COST $= \frac{1640.70}{.08} = 20,509$

$\boxed{B}$

7 $7000 = 1200 + 900(P/A, ?, 10)$

$(P/A, ?, 10) = 6.444$

ABOUT 9%

$\boxed{C}$

8 $200,000 + 200,000(P/F, 6\%, 5)$

$200,000(1 + .7473) = 349,460$

$\boxed{D}$

9 THE INITIAL BOOK VALUE IS 15,000

EACH YEAR, DEPRECIATION $= \frac{1}{10}(15,000 - 2800)$

$= 1220$

$\frac{1220}{15,000} = .081$

$\boxed{A}$

10 $P = -1000 + 200(P/F, 20\%, 8) - 15(P/A, 20\%, 8)$

$= -1000 + 200(.2326) - 15(3.8372)$

$= -1011$

$\boxed{D}$

11 SINCE THE TWO TERMS CONTAINING X ARE NEGATIVE, THE MAXIMUM PROFIT OCCURS WHEN $X = 0$.

$\boxed{A}$

12 $P = -8000 + 800(P/F, 10\%, 5)$
$\qquad\qquad - (1000 + 80)(P/A, 10\%, 5)$

$= -8000 + 800(.6209) - (1080)(3.7908)$

$= -11,597$

$\boxed{D}$

13 $-200,000 + (.8)(1,000,000)(5.75 - 1.53)$

$= 3,176,000$

$\boxed{B}$

14 $60,000 + .021N = 78,000 + .008N$

$.013N = 18,000$

$N = 1,384,615$

$C = 60,000 + .021(1,384,615)$

$= 89,077$

$\boxed{A}$

15 $.2(200) + .3(300) + .3(400) + .1(500) + .1(600)$

$= 360$

$\boxed{E}$

FLUID STATICS

1 ABSOLUTE VISCOSITY IS VERY TEMPERATURE DEPENDENT. (ABSOLUTE VISCOSITY AND DYNAMIC VISCOSITY ARE THE SAME).

$\boxed{E}$

2 AVERAGE DEPTH $= 2 + \frac{1}{2}(8) = 6$

AVERAGE PRESSURE $= (6)(62.4) = 374.4$ PSF

AREA $= (8)(8) = 64$ FT^2

FORCE $= (64)(374.4) = 23,962$ LBF

$\boxed{D}$

3 FOR A TOTALLY SUBMERGED OBJECT, THE CENTER OF MASS (C.G.) MUST BE BELOW THE CENTER OF BUOYANCY.

$\boxed{E}$

4 BUOYANT FORCE IN THE WATER = 65-42 = 23 LBF

VOLUME = $\frac{23}{62.4}$ = .369 FT^3

$SG = \frac{65}{(.369)(62.4)}$ = 2.82

$\boxed{D}$

5 BUOYANT FORCE = $\frac{(4500)(64.0)}{2000}$ = 144 TONS

$\boxed{E}$

6 A CONDENSER OPERATES IN A VACUUM.

$(14.7 - (24)(.491))(144)$ = 419.9 PSF

$\boxed{D}$

7 FIND THE WEIGHT OF WATER ABOVE THE SPHERE.

VOLUME OF CYLINDER OF RADIUS 2 AND HEIGHT 8

$\pi (2)^2 (8)$ = 100.5

VOLUME OF HEMISPHERE

$\frac{1}{2} \left(\frac{4}{3}\right) \pi (2)^3$ = 16.8

VOLUME OF WATER = 100.5 - 16.8 = 83.7

WEIGHT OF WATER = (83.7)(.0361) = 3.02

$\boxed{A}$

8 $\mu = \frac{v\rho}{g} = \frac{(20) \, FT^2 (.87)(62.4) \, LB \quad SEC^2}{SEC \qquad FT^3 (32.2) \, FT}$

$= .337 \frac{SEC-LB}{FT^2}$

BUT, A SLUG = $\frac{LB-SEC^2}{FT}$, SO

$\mu = .337 \frac{SLUG}{FT-SEC}$

$\boxed{A}$

9 A THREE-DIMENSIONAL OBJECT DOES NOT HAVE A CENTER OF PRESSURE.

$\boxed{C}$

10 FOR EXAMPLE, PRESSURE AT A DEPTH

$\Delta P = \frac{g}{g_c} \rho (z_1 - z_2)$

$g_c = \frac{g \rho \Delta z}{\Delta P} = \left(\frac{FT}{SEC^2}\right)\left(\frac{LBM}{FT^3}\right)(FT)\left(\frac{FT^2}{LBF}\right)$

$= \frac{LBM-FT}{SEC^2-LBF}$

$\boxed{E}$

11 $\bar{z} = 2 + \frac{1}{2}(6)$ = 5 FT

$\bar{P} = 5(62.4)$ = 312 PSF

$F = \bar{P}A = (312)(6)(4)$ = 7488 LBF

$\boxed{A}$

12 $h_c = 5$

$I_c = \frac{(4)(6)^3}{12}$ = 72

$A = 24$

$h_R = 5 + \frac{72}{(24)(5)}$ = 5.6

$\boxed{E}$

13 $p = (300) \, FT \, (12) \, ^{IN}/_{FT} \, (.0361) \, ^{LB}/_{IN^3}$ = 130 psi

$\sigma_{HOOP} = \frac{Pr}{t}$

$t = \frac{(130)(24)}{18,000}$ = .173 IN

$\boxed{C}$

14 $\sigma_{HOOP} = \frac{Pr}{t}$

$\sigma_{LONG} = \frac{Pr}{2t}$

THESE ARE BOTH PRINCIPAL STRESSES. THEY DO NOT COMBINE

$\boxed{E}$

15 $(29.92 - 20)(.491)$ = 4.87 PSI

$\boxed{A}$

FLUID DYNAMICS

1 $N_{Re} = \frac{Dv\rho}{\mu g} = \frac{DG}{\mu}$

$G = \frac{v\rho}{g}$ = MASS FLOW RATE PER UNIT AREA

$\boxed{E}$

2

$\boxed{D}$

3 AREA IN FLOW = $\left(\frac{1}{2}\right)(2)(4)$ = 4 FT^2

WETTED PERIMETER = 1+4+1 = 6 FT

$r_H = \frac{4}{6}$ = .67

$\boxed{E}$

4 $A_1 v_1 = A_2 v_2 \longrightarrow \left(\frac{\pi}{4}\right)(1)^2 (10) = \left(\frac{\pi}{4}\right)(2)^2 v_2$

$v_2 = 10/4$ = 2.5

$\boxed{B}$

5 ASSUME $C_F = 1$

$$F_{VA} = \frac{1}{\sqrt{1 - (1/3)^2}} = 1.06$$

$$\Delta P = \frac{(V_2/F_{VA})^2}{2g} = \frac{(40/1.06)^2}{(2)(32.2)} = 22.11$$

$$\boxed{C}$$

6
$$\boxed{C}$$

7 $h_f = K\left(\frac{V^2}{2g}\right)$

$\frac{V^2}{2g}$ IS THE VELOCITY (DYNAMIC) HEAD

$$\boxed{B}$$

8 $\dot{w} = (2)\frac{FT^3}{SEC}(62.4)\frac{LB}{FT^3} = 124.8\ LB/SEC$

$$HP = \frac{\dot{w}\Delta z}{550} = \frac{(124.8)(50)}{550} = 11.35$$

$$\boxed{B}$$

9 $h_f = \frac{fLv^2}{2Dg}$ or $f = \frac{2Dg\,h_f}{Lv^2}$

$$f = \frac{(2)(9/12)(32.2)(26.2)}{(1000)(10)^2} = .01687$$

$$\boxed{A}$$

10 $N_{Re} = \frac{Dv\rho}{\mu g} = \frac{(1)(10)(1.22)(62.4)}{(.00122)(32.2)} = 19379$

$$\boxed{D}$$

11 USE THE BERNOULLI EQUATION

$$\frac{(16.8)(144)}{62.4} + \frac{v^2}{(2)(32.2)} = \frac{(17.2)(144)}{62.4} + \frac{(6.2)^2}{(2)(32.2)}$$

$$v = 9.89$$

$$\boxed{B}$$

12 AT CRITICAL FLOW, $\frac{v^2}{g} = d$, SO

$$d = \frac{(2)^2}{32.2} = .124$$

$$\boxed{A}$$

13 FLOW WORK AND PV-WORK ARE THE SAME THING.
THE VELOCITY TERM IS NOT USED.

$$W = PV = (80)(144)(1.5) = 17,280\ FT\text{-}LB$$

$$\boxed{A}$$

14 POWER AND SPEED ARE RELATED BY

$$\frac{P_1}{P_2} = \frac{(N_1)^3}{(N_2)^3}$$

SO, $P_2 = P_1\left(\frac{750}{720}\right)^3 = 1.13 P_1$

$$\boxed{A}$$

15 $V = \frac{Q}{A} = \frac{(.139)(144)}{1} = 20\ FT/SEC$

$$F = -\left[\frac{(.139)(62.4)}{32.2}\right](20)\left[1 - (.5)\right]$$

$$= 8.08\ LBF$$

$$\boxed{D}$$

THERMODYNAMICS

1

$Q = Tds$

T vs S graph: CONSTANT P OR CONSTANT T

$$\boxed{E}$$

2 REVERSIBLE ADIABATIC IS THE SAME AS ISENTROPIC.

$C_v\,dT = du$

$p\,dV = dW$

$$\boxed{A}$$

3
$$\boxed{E}$$

4 $W = P(V_2 - V_1)$

$$= \frac{(15.3 + 14.7)(144)(20)}{778} = 111.0$$

$$\boxed{B}$$

5 $V_2 = V_1\left(\frac{P_1}{P_2}\right)^{1/k} = 2\left(\frac{14.7}{30}\right)^{1/1.4} = 1.20$

$$\boxed{A}$$

6 ENERGY IS REQUIRED TO RAISE WATER AND
STEAM TEMPERATURES AND TO PROVIDE HEAT
OF VAPORIZATION

$$g = 2\left[1 + 1 + 970.3 + .5\right] = 1945.6$$

CLOSEST ANSWER IS E.

(USE OF STEAM TABLES GIVES 1938)

$$\boxed{E}$$

7 THE PERCOLATOR GIVES OFF BOTH STEAM AND HEAT.

$$\boxed{B}$$

8

$$\boxed{E}$$

9 STRICTLY SPEAKING, NONE ARE CORRECT. BUT, $\Delta U = -W$, SO THE BEST RESPONSE IS

$$\boxed{B}$$

10 $M = \dfrac{PV}{RT} = \dfrac{(250)(144)(30)}{\left(\dfrac{1545}{24}\right)(460+90)} = 30.5$

$$\boxed{B}$$

11 $T_1 = 460 + 68 = 528$

$T_2 = (528)\left(\dfrac{75}{15}\right)^{\frac{1.4-1}{1.4}} = 836.3$

$C_p = \dfrac{R^* k}{k-1}$

$C_v = C_p - R^* = \dfrac{R^*}{k-1} = 3862.5 \ \dfrac{\text{ft-lbf}}{\text{pmole-}^\circ R}$

$W = mC_v(T_1 - T_2) = (1)(3862.5)(528-836.3)$
$= -1.19 \ EE6 \ \dfrac{FT\text{-}LBF}{pmole}$

$$\boxed{C}$$

12 $ds = M\int \dfrac{d\phi}{T} = M\int \dfrac{C_p\, dT}{T}$

$\Delta S = MC_p\left[\ln T\right]_{560}^{760} = (50)(.0915)\left[\ln(760) - \ln(560)\right]$

$= 1.397$ FOR THE HEATING BLOCK

SIMILARLY FOR THE COOLING BLOCK,

$\Delta S = (50)(.0915)\left[\ln(760) - \ln(960)\right]$

$= -1.069$

$\Delta S_{total} = 1.397 - 1.069 = .328 \ BTU/^\circ R$

$$\boxed{D}$$

13 $h = 298.0 + .97(888.8) = 1160.5$

$$\boxed{E}$$

14 $P_2 = P_1\left(V_1/V_2\right) = 50\left(\dfrac{10}{1}\right) = 500 \ PSIA$

$$\boxed{E}$$

15 SINCE $h = \mu + pV$

$\Delta h = \Delta \mu + p\Delta V$

FROM THE STEAM TABLES FOR $32^\circ F$,

$p\Delta V = \dfrac{(.08854)(144)(3306)}{778} = 54.2$

$$\boxed{D}$$

POWER CYCLES

1 ISOCHORIC IS THE SAME AS CONSTANT VOLUME

$$\boxed{A}$$

2 $\eta = \dfrac{(900+460)-(100+460)}{900+460} = .588$

$$\boxed{C}$$

3 $COP = \dfrac{15+460}{(75+460)-(15+460)} = 7.92$

$$\boxed{D}$$

4 $COP = \dfrac{460+120}{(460+120)-(460+10)} = 5.27$

$$\boxed{C}$$

5 $\eta = \dfrac{OUTPUT}{INPUT} = \dfrac{1 \ HP}{(.4)(18,500) \ BTU/HR}$

$= \dfrac{(1)\ HP\ (550)\frac{ft\text{-}lb}{HP\text{-}SEC}}{\dfrac{(.4)(18500)}{3600}(778)\frac{FT\text{-}LB}{BTU}} = .344$

$$\boxed{A}$$

6 $W_{NET} = Q_{NET} = T_{HIGH}\Delta S_{a-b} - T_{LOW}\Delta S_{c-d}$

$= 860(.9608) - 660(.9608) = 192.2$

$$\boxed{B}$$

7 $V_1 = \dfrac{6000}{60} = 100 \ FPS$

$V_2 = \dfrac{25000}{60} = 416.6 \ FPS$

$\Delta E = h_1 - h_2 + \dfrac{V_1^2 - V_2^2}{2g}$

$= (1200-900) + \dfrac{(100)^2 - (416.6)^2}{(2)(32.2)(778)} = 296.7 \ \dfrac{BTU}{LOM}$

$W = (296.7)\left(\dfrac{25000}{3600}\right) - \dfrac{125000}{3600}$

$= 2025.7 \ BTU/SEC$

{MORE}

#7 CONTINUED

$$HP = \frac{(2025.7)(778)}{550} = 2865.4$$

E

8 FROM THE AIR TABLE,

$$h_1 = 119.48 \quad and \quad h_2 = 143.47$$

$$\Delta h = 24 \; BTU/LBM$$

D

9 FROM THE AIR TABLE,

$$h_1 = 255.96 \qquad h_2 = 157.92$$

$$\Delta h = 98.04 \; BTU/LBM$$

$$\Delta E = 2(98.04) = 196$$

THIS ASSUMES THE EXPANSION IS STEADY FLOW. IF IT IS ASSUMED THAT THE PROCESS IS CLOSED, THEN THE ANSWER IS AROUND 135 TO 140 BTU.

A

10 $1192 - 1158 = 34 \; BTU/LBM$

B

11

C

12

D

13

D

14

D

15

B

CHEMISTRY

1 MW OF $H_2SO_4 = 2(1) + 32.1 + 4(16) = 98.1$

$$M = \frac{588/98.1}{2} = 3$$

C

2 THE OXYGEN IN (A) DOES NOT BALANCE

A

3

E

4

D

5 $3 OH^- + PO_4^{+++} \longrightarrow PO_4(OH)_3$

$\frac{1}{3}$ IS THE RATIO

A

6 $CH_4 + 2O_2 \longrightarrow 2H_2O + CO_2$

$$1(-17.9) + 2(0) \longrightarrow 2(-57.8) + (-94.1)$$

$$\Delta E = 2(-57.8) + (-94.1) - (-17.9) = -191.8$$

C

7 $K_2Cr_2O_7 + 14 HCl \longrightarrow 2KCl + 2CrCl_3 \cdot$
$$+ 7H_2O + 3Cl_2$$

A

8 ZN BECOMES MORE POSITIVE (LESS NEGATIVE) AND SO, IT IS OXIDIZED.

H BECOMES MORE NEGATIVE, AND IS REDUCED.

B

9 C_2H_5OH IS ETHYL ALCOHOL

A

10

A

11 $HC_2H_3O_2 + H_2O \longrightarrow H_3O^+ + C_2H_3O_2^-$

$$K = 1.7 \; EE\text{-}5 = \frac{[H_3O^+][C_2H_3O_2^-]}{[HC_2H_3O_2]}$$

BUT $[H_3O^+] = [C_2H_3O_2^-]$

SO,
$$1.7 \; EE\text{-}5 = \frac{[H_3O^+]^2}{[HC_2H_3O_2]}$$

NOW, $[H_3O^+] = [x]M = .03x$
(X IS THE FRACTION IONIZED)

SO
$$[HC_2H_3O_2] = (1-x)(.03)$$

ASSUME X IS SMALL SO THAT $1-x \approx 1$

$$1.7 \; EE\text{-}5 = \frac{x^2 (.03)^2}{(1)(.03)}$$

$$X = .0238$$

B

12 ZnO IS 80.3% Zn BY WEIGHT, SO
$$(.803)(16.28) = 13.08 \; g \; Zn$$

CO_2 IS 27.3% C BY WEIGHT, SO
$$(.273)(35.2) = 9.6 \; g \; C$$

H_2O IS 11.1% H BY WEIGHT, SO
$$(.111)(18) = 2 \; g \; H$$

$$\left. \begin{array}{l} \frac{13.08}{65.38} = .2 \\[4pt] \frac{9.6}{12} = .8 \\[4pt] \frac{2}{2} = 1 \end{array} \right\} \quad Z_{.2}C_{.8}H_1 \longrightarrow ZC_4H_5$$

E

13 $\quad$ $\boxed{A}$

14 $\quad (1.5)(.3)(342.3) = 154$
$\quad \boxed{A}$

15 $\quad (23.05)(.1) = (10)(x)$
$\qquad x = .2305$
$\qquad\qquad \boxed{D}$

STATICS

1 $\quad \bar{y} = \dfrac{(8)(4) + (8)(9)}{8+8} = 6.5$
$\qquad\qquad \boxed{B}$

2 $\quad \dfrac{(8)(1)^3}{12} + \dfrac{(2)(4)^3}{12} = 11.33$
$\qquad\qquad \boxed{D}$

3 $\quad r = \sqrt{I/A} = \sqrt{\dfrac{11.33}{16}} = .84$
$\qquad\qquad \boxed{D}$

4 $\quad M = Pd$
$\qquad\qquad \boxed{C}$

5 $\quad \sum M_L = 12(R) - 10(d) - (5+d)5 = 0$
$\quad$ BUT $L = R = {}^{15}/_2$
$\quad$ SOLVING FOR d, $\quad d = 4.33$
$\qquad\qquad \boxed{A}$

6 $\quad \sqrt{(6)^2 + (2)^2 + (10)^2} = 11.83$
$\qquad\qquad \boxed{D}$

7 $\quad 300 \#$
$\qquad\qquad \boxed{A}$

8 $\quad \sum M_B = 18(A) - 48(12) = 0$
$\qquad\qquad A = 32$
$\qquad\qquad \boxed{C}$

9 $\quad$ LENGTH $AB = \dfrac{10}{TAN\,60°} = 5.77$

$\quad \sum M_B \circlearrowleft : (86)(8)\cos 60° - B\cos 30°(10)$
$\qquad\qquad\qquad - B\sin 30°(5.77) = 0$
$\qquad\qquad B = 29.78$
$\quad |AB| = B_x = (29.78)(\cos 30°) = 25.8$
$\qquad\qquad \boxed{E}$

10 $\quad \sum F_x = 100 + 300(\cos 30°) = 359.8$
$\quad \sum F_y = 500 - 300(\sin 30°) = 350$
$\qquad R = \sqrt{(359.8)^2 + (350)^2} = 501.95$
$\qquad\qquad \boxed{B}$

11 $\quad T_y = 50$
$\quad T_x = 50(TAN\,20°) = 18.2$
$\qquad\qquad \boxed{A}$

12 $\quad F = 6.25 + 18.75 = 25$
$\qquad 25(.25) = 6.25$
$\qquad 75(.25) = 18.75$
$\qquad\qquad \boxed{B}$

13 $\quad$ N = NORMAL FORCE
$\qquad = (\tfrac{4}{5})800 = 640$
$\quad$ H = HORIZONTAL SLIDING FORCE
$\qquad = (\tfrac{3}{5})800 = 480$

$\quad$ THE AVAILABLE DYNAMIC FRICTION =
$\qquad (.20)(640) = 128$

$\quad$ MORE FRICTION IS AVAILABLE FOR THE STATIC CASE. SINCE H AND THE APPLIED FORCE ARE OUT OF BALANCE BY $480 - 400 = 80$, AND SINCE THE AVAILABLE FRICTION IS 128, THERE IS NO SLIDING, AND
$\qquad F_6 = 80 \nearrow$ UPHILL TO OPPOSE MOTION
$\qquad\qquad \boxed{D}$

14 $\quad \sum M_c : A(4.42 + 3.58) - 600(3.58) = 0$
$\qquad\qquad A = 268.5$
$\qquad\qquad \boxed{B}$

15 $\quad AB = \dfrac{5}{3}(100) = 166.6$
$\qquad\qquad \boxed{C}$

DYNAMICS

1 $\quad$ USE THE IMPULSE-MOMENTUM PRINCIPLE
$\qquad Tt = J\,\Delta\omega$
$\qquad T = 2F, \quad t = 5, \quad \Delta\omega = 8$
$\qquad J = \tfrac{1}{2}(1)(2)^2 = 2$
$\qquad (2F)(5) = (2)(8) \longrightarrow F = 1.6$
$\qquad\qquad \boxed{B}$

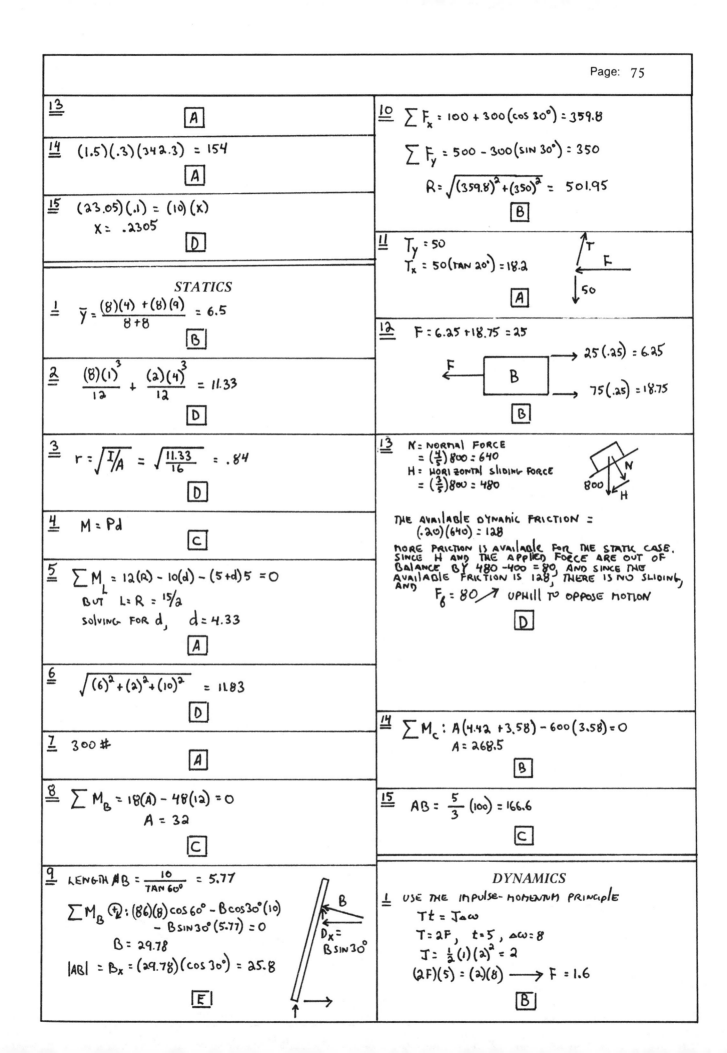

2
$F_y = F_x = 70.7$
$F_{friction} = .20(100 - 70.7) = 5.86$
$a_x = F/m = \dfrac{(70.7 - 5.86)}{\frac{100}{32.2}} = 20.88 \ ft/sec^2$

[D]

3
$\Delta E_{potential} = (10)(\sin 30°)(100) = 500 \ ft\text{-}lb$
$\Delta E_{kinetic} = (\frac{1}{2})(\frac{10}{32.2})(52)^2 = 419.9$
$E_{friction} = 500 - 419.9 = 80.1$

[A]

4
$\dfrac{(\frac{1}{2})(4)(4) + (\frac{1}{2})(4)(2)}{6} = 2$

[B]

5
$S = V_0 t + \frac{1}{2}at^2$
$[0,2] \ S = 0 + (\frac{1}{2})(1)(2)^2 = 2$
$V(2) = at = (1)(2) = 2$
$[2,3] \ S = (2)(1) + \frac{1}{2}(3)(1)^2 = 3.5$
$S_{0,3} = 2 + 3.5 = 5.5$ [D]

6
SINCE THE ANGLE IS GREATER THAN ARCTAN(.3), $F = 0$

[A]

7
CHOICE E IS THE ONLY ONE FOR WHICH $g = g$ AT $h = 0$.

[E]

8
$\Delta E_P = \Delta E_K$
$(5)(1) = \frac{1}{2}(\frac{5}{32.2})(v)^2$
$V = 8.02$

[B]

9
$45 \ MPH = 66 \ FPS$
$W = (2200)(66)(.03) = 4356 \ \dfrac{ft\text{-}lb}{sec}$
$hp = \dfrac{4356}{550} = 7.92$

[B]

10
$\frac{d}{dV}(\frac{1}{2}hv^2) = hV$

[D]

11
$160 = \dfrac{210}{32.2}(32.2 - a)$
$a = 7.66$

[C]

12
$V = \frac{ds}{dt} = 3t^2 - 14t - 13$
$V = 0$ AT $t = 5.45$
[E]

13
TO CONSERVE MOMENTUM,
$45(2) = 40(1.2) + 45(x)$
$X = .93 \ FT/SEC$ TO THE RIGHT
$e = \dfrac{.93 - 1.2}{0 - 2} = .135$

[B]

14
$V_0 = 88 \ FT/SEC$
$V_f = \sqrt{(88)^2 + (2)(3)(300)} = 97.7 \ ft/sec$
$a_N = \frac{v^2}{r} = \dfrac{(97.7)^2}{400} = 23.86$
$a_{total} = \sqrt{(23.86)^2 + (3)^2} = 24.05$

[A]

15
$y = (V_0 \sin \theta)t - \frac{1}{2}gt^2$
$2000 = (4000)\sin 30° t - (.5)(32.2)t^2$
SO, $t = 123.2 \ SEC$
$X = (4000)(\cos 30°)(123.2) = 426,777 \ ft$

[B]

MECHANICS OF MATERIALS

1

[A]

2

[C]

3
$F_{critical} = \dfrac{\pi^2 EI}{L^2} = \dfrac{(k\pi)^2 EA}{L^2}$

[E]

4
$I = \dfrac{(2)(2)^3}{12} = 1.33$
$\sigma = \dfrac{My}{I} = \dfrac{(1030)(12)(.5)}{1.33} = 4635 \ psi$

[C]

5
$20/6 = 3.33$

KERNAL

[C]

6 FIRST, SOLVE FOR X
$.25 = \dfrac{x^4}{1.14 \ E9}$
$X = 129.93$
{MORE}

#6 CONTINUED

NOW, DIFFERENTIATE FOR y'

$$y'(x) = \frac{4x^3}{1.14\,EE9}$$

$$y'(129.93) = .0077$$

$$\boxed{B}$$

7

$$I_{MIN} = \frac{(2)(1)^3}{12} = .1667$$

$$A = (2)(1) = 2$$

$$K_{MIN} = \sqrt{\frac{.1667}{2}} = .289$$

ASSUME $\frac{L}{K} < 100$, SO

$$L = 28.9 \quad \text{(PINNED ENDS)}$$
$$L = 2(28.9) = 57.8 \quad \text{(FIXED ENDS}$$

$$\boxed{B}$$

8

$$\sigma_{HOOP} = \frac{Pr}{t}$$

$$12,000 = \frac{P(12)}{.125} \longrightarrow P = 125 \; PSI$$

PRESSURE IN THE ENDS
CAN BE TWICE THIS BUT
HOOP STRESS CONTROLS.

$$\boxed{E}$$

9

$$A_{SHEAR} = \frac{1}{2}(\pi)\left(\frac{1}{2}\right) = .7854$$

$$F = (43,000)(.7854) = 33,772 \; LB$$

$$\boxed{C}$$

10

$$\frac{6}{9}(50,000) = 33,333$$

$$\boxed{B}$$

11

$$\delta = \frac{FL}{AE} = \frac{(50)(4)(12)}{(.5)(3EE7)} = 1.6\,EE\text{-}4$$

$$\boxed{B}$$

12

$$y = \sqrt{(72.36)^2 - (72)^2} = 7.21$$

VERTICAL CABLE ELONGATION
$= (10)(12)(.008) = .96$

TOTAL $= 7.21 + .96 = \underline{8.17}$

$$\boxed{D}$$

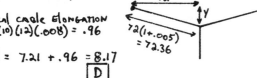

$$72(1 + .005)$$
$$= 72.36$$

13

$$\tau = \frac{Tc}{J} \qquad \tau = 8000 \quad T = \text{TORQUE} \quad c = .5$$
$$J = \left(\frac{1}{2}\right)\pi(.5)^4 = .0982$$

$$T = \frac{(8000)(.0982)}{.5} = 1570\; IN\text{-}LBF$$

$$\boxed{C}$$

14

$$315 \; RPM = 5.25 \; RPS$$
$$W = (100)(550)(12) = 6.6\,EE5 \; \frac{IN\text{-}LB}{SEC}$$
$$V = (5.25)(\pi)(5) = 82.47 \; IN/SEC$$
$$F = \frac{W}{V} = 8\,EE3$$

$$T = Fr = (8\,EE3)(2.5) = 2\,EE4$$

$$A_{BOLT} = \frac{\pi}{4}\left(\frac{3}{8}\right)^2 = .110 \; IN^2$$

$$J = x\left[(.110)(2.5)^2\right] = (.6875)x$$

$$\tau = \frac{Tc}{J}$$

$$10,000 = \frac{(2\,EE4)(2.5)}{(.6875)x}$$

$$x = 7.25 \quad (SAY\;8)$$

$$\boxed{C}$$

15

AXIAL DEFORMATION $= \dfrac{FL}{AE}$

$$= \frac{(10,000)(10)}{\left(\frac{\pi}{4}\right)(1)^2(1\,EE7)} = 1.27\,EE\text{-}2$$

AXIAL STRAIN $= 1.27\,EE\text{-}2/10 = 1.27\,EE\text{-}3$

DIAMETRAL DEFORMATION $=$
$$.33(1.27\,EE\text{-}3) = 4.2\,EE\text{-}4$$

$$\boxed{A}$$

D.C. ELECTRICITY

1

$$R_t = R_1 + \frac{R_2 R_3}{R_2 + R_3} = \frac{R_1 R_3 + R_1 R_2 + R_2 R_3}{R_2 + R_3}$$

$$I_t = \frac{V}{R_t}$$

$$I_{R_2} = \left(\frac{R_3}{R_2 + R_3}\right)I_t$$

$$\boxed{B}$$

2

USE SUPERPOSITION AND SYMMETRY. DUE
TO THE LEFT-HAND SOURCE ALONE

$$I_t = \frac{10\;V}{1 + \frac{(1)(1)}{1+1}} = 6.67$$

$$I_{R_3} = \frac{1}{2}(6.67) = 3.33$$

DUE TO SYMMETRY, $2(3.33) = 6.67$

$$\boxed{B}$$

3 $(2)(x)(y)$ volt-coulombs $= 2xy$ JOULES

THIS MAY BE POSITIVE OR NEGATIVE DEPENDING ON THE DEFINITION OF "SYSTEM"

$\boxed{A}$ $\boxed{C}$

4 $R_t = \frac{(4)(6)}{4+6} = 2.4$

$I_t = \frac{20}{2.4} = 8.33$

$I = \left(\frac{6}{4+6}\right) 8.33 = 5.0$

$\boxed{D}$

5 $\boxed{E}$

6 $E = \frac{q}{4\pi\epsilon r^2}$ $\boxed{D}$

7 $E = \frac{q}{4\pi\epsilon (R+r)^2}$ $\boxed{E}$

8 FLUX DECREASES IN PROPORTION TO THE AREA

$\boxed{D}$

9 $\frac{(1)(1)}{4\pi\epsilon r^2} = \frac{(1)(4)}{4\pi\epsilon(1.5-r)^2}$

$\frac{1}{r^2} = \frac{4}{(1.5-r)^2} \longrightarrow r = .5$

$\boxed{A}$

10 $3-2 = 1\,dB$

$1 = 10 \; LOG_{10}\left(\frac{P_2}{P_1}\right)$

$\quad = 10 \; LOG_{10}\left(\frac{P_2}{.001}\right)$

$P_2 = .00125\,W$

$\boxed{B}$

11 $I_c^* = 3(1.8) = 5.4\,A$

$(4.05)(R_c) = (11-4.05)R_s$

$R_s = .58 R_c$

$(.75)(5.4) = 4.05$

$\boxed{C}$

12 $\frac{3000}{(5)(60)} = 10\,A$

$\boxed{D}$

13 $\frac{(3)(.7457)(24)(.10)}{.95} = \$5.65$

$\boxed{E}$

14 $\alpha = \frac{25.9-20.3}{(20.3)(50-20)} = .009195 \; @ \; 20.3°$

α IS AROUND .008 OVER THE AVERAGE INTERVAL.

$\boxed{A}$

15 $R = \frac{L}{\mu A} = \frac{(5)\pi \quad (.0254)\,{}^M/_{IN}}{(3.2)(\frac{\pi}{4})(1)^2(.0254)^2\,(^M/_{IN})^2}$

$\quad = 246.1$

$(2)(2000) = \phi(246.1)$

$\phi = 16.26$

$\boxed{A}$

A.C. ELECTRICITY

1 $a = \frac{N_1}{N_2}$, $a^2 = \frac{Z_P}{Z_S} = \frac{3200}{8}$

$a = 20$ $\boxed{C}$

2 $a = \frac{N_1}{N_2} = \frac{1}{2} = \frac{I_2}{I_1}$

SO $I_1 = 2 I_2$

$\boxed{D}$

3 $I_{eff} = (.707)(150) = 106.5$

$\boxed{D}$

4 $|Z| = \sqrt{(5)^2 + (2)^2} = 5.39$

$\phi = ARCTAN\left(\frac{-2}{5}\right) = -21.8$

SO, $Z = 5.39 \angle -21.8°$

$Y = \frac{1}{Z} = .186 \angle 21.8°$

$\boxed{E}$

5 $.2 = 1 - e^{\frac{-t}{(3)(1)}}$

$t = .67$ $\boxed{B}$

6 $V(t) = L\frac{dI}{dt} \longrightarrow 30t = 3\frac{dI}{dt}$

$\frac{10}{2}t^2 = I$ $\boxed{E}$

7 $Z = \dfrac{V}{I} = \dfrac{120}{.5} = 240 \,\Omega$

$240 = \sqrt{(50)^2 + (X_c)^2}$

$X_c = 234.7 \,\Omega$

$234.7 = \dfrac{1}{2\pi(60)C}$

$C = 11.3 \, EE-6 \, F$

$\boxed{B}$

---

8 FOR THE LEFT SET

$\dfrac{1}{Z_L} = \dfrac{1}{4\angle 0} + \dfrac{1}{6\angle 90} = .25\angle 0 + .1667\angle{-90}$

$= .3\angle{-33.7}$

SO, $Z_L = 3.33\angle{33.7} = 2.77 + J1.85$

FOR THE RIGHT SET

$\dfrac{1}{Z_R} = \dfrac{1}{2\angle 0} + \dfrac{1}{3\angle{-90}} = .5\angle 0 + .333\angle 90$

$= .6\angle{33.67}$

SO, $Z_R = 1.667\angle{-33.7} = 1.39 - J.925$

$Z_{total} = (2.77 + 1.39) + J(1.85 - .925)$

$= 4.16 + J.925$

$= 4.26\angle{12.53}$

$\boxed{B}$

---

9 $V_{phase} = \dfrac{440}{\sqrt{3}} = 254$

$P = (254)(25)(.60) = 3810 \, WATTS$

$\boxed{E}$

---

10 $\boxed{D}$

---

11 $I = \dfrac{P}{V} = \dfrac{30,000}{120} = 250 \, AMP \, RMS$

$\boxed{D}$

---

12 $Z = \sqrt{(10)^2 + (24)^2} = 26$

$\tan\phi = \dfrac{-24}{10} \qquad \phi = -67.4$

$26\angle{-67.4} = 26\left[e^{-J1.18}\right]$

$\boxed{A}$

---

13 TOTAL VOLTAGE ACROSS THE RESISTOR IS

$V = \dfrac{Q_1}{C_1} + \dfrac{Q_2}{C_2} = \dfrac{600}{6} + \dfrac{0}{8} = 100V$

$I = \dfrac{V}{R} = \dfrac{100}{2} = 50 A$

$\boxed{B}$

---

14 BEFORE

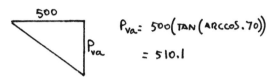

$P_{va} = 500(TAN(ARCCOS.70))$

$= 510.1$

AFTER

$P_{va} = 650(TAN(ARCCOS.85))$

$= 402.8$

MOTOR

$510.1 - 402.8 = 107.3$

$COS\left[ARCTAN\left(\dfrac{107.3}{150}\right)\right] = .81$

$\boxed{C}$

---

15 $X_L = 2\pi Lf = 2\pi(1)(60) = 377 \,\Omega$

ASSUME 2 AMPS IS RMS VALUE

$P = I^2X = (2)^2(377) = 1508 \, va$

$\boxed{C}$

MATERIALS SCIENCE

1 TOUGHNESS IS THE ENERGY REQUIRED TO FRACTURE A UNIT VOLUME. THIS CORRESPONDS TO THE AREA UNDER THE STRESS-STRAIN CURVE

$E = \tfrac{1}{2}(.0015)(40,000) + (40,000)(.0025 - .0015)$

$= 70 \dfrac{IN-LB}{IN3} = 70 \, PSI$

$\boxed{D}$

2 $\boxed{E}$

3 $\boxed{C}$

4 ONE IN THE CENTER AND $\tfrac{1}{8}$ OF EACH OF EIGHT CORNERS $= 2$

$\boxed{B}$

5 $\boxed{C}$

6 $\boxed{C}$

7 ATOMIC WEIGHT OF ALUMINUM IS 27 g/gmole.
1 CC WEIGHS 2.7 g WHICH IS 1/10 OF A mole
$(.1)(6.023\ EE23) = 6.023\ EE22$

$\boxed{C}$

8 $\boxed{B}$

9 $\boxed{B}$

10 $\boxed{B}$

11 $\boxed{E}$

12 $\boxed{A}$

13 $\boxed{E}$

14 $\boxed{D}$

15 $\boxed{C}$

PERIPHERAL SCIENCES

1 $1000 = \dfrac{(2)(3)(k)(460)}{1}$

$K = .362$

$\boxed{C}$

2 $\boxed{A}$

3 $\boxed{A}$

4 $\boxed{B}$

5 BOTH C AND D COULD BE ARGUED. PROBABLY, D IS BEST

$\boxed{D}$

6
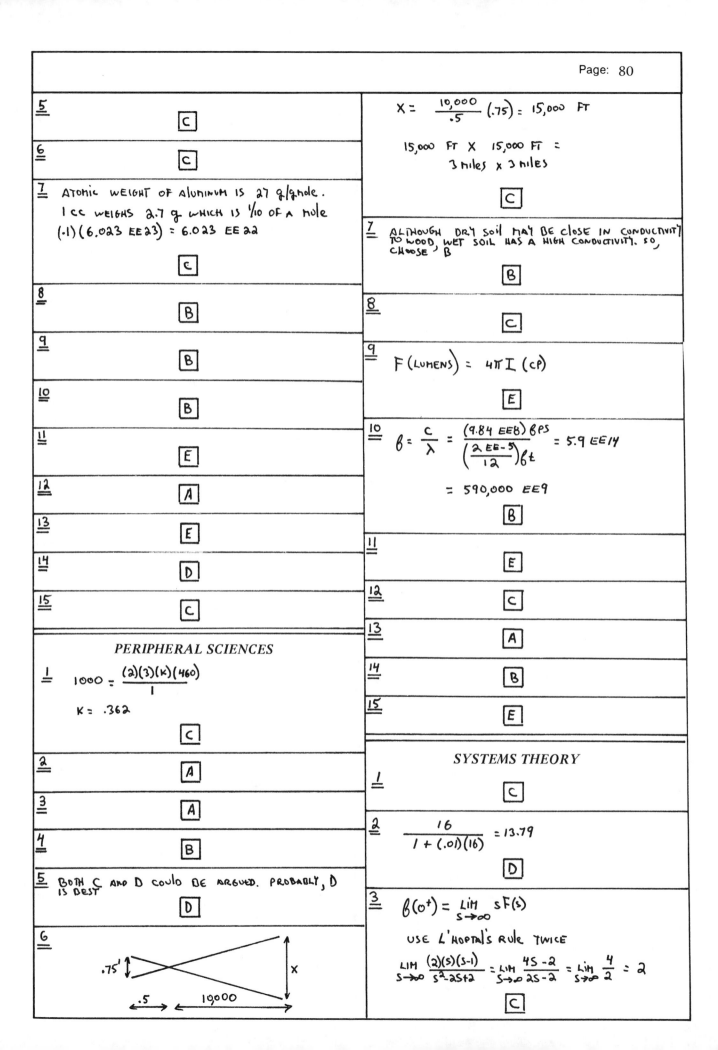

$X = \dfrac{10,000}{.5}(.75) = 15,000\ FT$

$15,000\ FT \times 15,000\ FT = $
3 miles x 3 miles

$\boxed{C}$

7 ALTHOUGH DRY SOIL MAY BE CLOSE IN CONDUCTIVITY
TO WOOD, WET SOIL HAS A HIGH CONDUCTIVITY. SO,
CHOOSE B

$\boxed{B}$

8 $\boxed{C}$

9 $F\ (LUMENS) = 4\pi I\ (cP)$

$\boxed{E}$

10 $\beta = \dfrac{c}{\lambda} = \dfrac{(9.84\ EE8)\ BPS}{\left(\dfrac{2\ EE-5}{12}\right)6t} = 5.9\ EE14$

$= 590,000\ EE9$

$\boxed{B}$

11 $\boxed{E}$

12 $\boxed{C}$

13 $\boxed{A}$

14 $\boxed{B}$

15 $\boxed{E}$

SYSTEMS THEORY

1 $\boxed{C}$

2 $\dfrac{16}{1+(.01)(16)} = 13.79$

$\boxed{D}$

3 $\beta(o^+) = \lim_{s\to\infty} sF(s)$

USE L'HOPITAL'S RULE TWICE

$\lim_{s\to\infty}\dfrac{(2)(s)(s-1)}{s^2-2s+2} = \lim_{s\to\infty}\dfrac{4s-2}{2s-2} = \lim_{s\to\infty}\dfrac{4}{2} = 2$

$\boxed{C}$

4 $S=0$ AND $S=3$ MAKE THE DENOMINATOR ZERO.

$$\boxed{C}$$

5 $f(\infty) = \lim_{s \to 0} sF(s) = 0$ SINCE S DOES NOT FACTOR OUT OF THE DENOMINATOR

$$\boxed{A}$$

6 $S=-4$ MAKES THE NUMERATOR ZERO

$$\boxed{C}$$

7
$$\boxed{E}$$

8 $(20)(100)(.5)(.01)(1.0) = 10$
$$\boxed{E}$$

9
$$\boxed{C}$$

10 (A) THROUGH (D) ARE NOT TRUE.
$$\boxed{E}$$

11

$$\frac{D}{A} + \frac{B}{C} = \frac{DC + BA}{AC}$$

SO, $G_6 = \dfrac{AC}{1 + (AC)\left[\dfrac{DC + BA}{AC}\right]}$

$$= \frac{AC}{1 + DC + BA}$$

$$\boxed{A}$$

12
$$\boxed{A}$$

13
$$\boxed{E}$$

14
$$\boxed{D}$$

15 $f(t) = f(-t) \longrightarrow$ EVEN FUNCTION SYMMETRY

$f(t) = -f(t+\pi) \longrightarrow$ ROTATIONAL SYMMETRY

TOGETHER, THESE MAKE UP QUARTER WAVE SYMMETRY.

$$\boxed{D}$$

COMPUTER SCIENCE

1
$$\boxed{B}$$

2
$(0011)_2$	$(0111)_2$	$(1101)_2$
$(3)_{10}$	$(7)_{10}$	$(13)_{10}$
$(3)_{16}$	$(7)_{16}$	$(D)_{16}$

$$\boxed{E}$$

3
$$\boxed{A}$$

4 NEGLECTING THE PARITY BIT, $(8)(12) = 96$
$$\boxed{E}$$

5
$$\boxed{C}$$

6
$$\boxed{A}$$

7 $\dfrac{(14.7)(3)}{(4)(2)} = 5.51$

THIS IS TRUNCATED TO THE INTEGER VALUE 5

$$\boxed{E}$$

8

$$\frac{75}{4} = 18 \text{ REMAINDER } 3$$

$$\frac{18}{4} = 4 \text{ REMAINDER } 2$$

$$\frac{4}{4} = 1 \text{ REMAINDER } 0$$

$$\frac{1}{4} = 0 \text{ REMAINDER } 1$$

$$(1023)_4$$

E

9

$$V_{OUT} = (-2)\left(\frac{20}{10}\right) = -4$$

C

10

E

11

E

12

J = 59432 (INTEGER)

SINCE J HAS MORE THAN 4 DIGITS IT WILL OVERFLOW THE AVAILABLE SPACE AND APPEAR AS ****

E

13 BASIC HAS NOT BEEN STANDARDIZED

A

14

D

15

$$(1.3)^{6.1} = 4.955$$

$$\frac{(5.7)(2.1)}{9.9} = 1.209$$

$$1.209 + 4.955 = 6.16$$

B

NUCLEONICS

1

$$1 \text{ WATT} = 1 \frac{JOULE}{SEC}$$

$$\frac{1}{(180 \text{ EE6})(1.6 \text{ EE-19})} = 3.47 \text{ EE10}$$

B

2 PROTONS AND ELECTRONS

B

3

A

4 THE ATOMIC WEIGHT OF IRON IS 56. THE VOLUME OF 1 GMOLE (56g) IS

$$\frac{56}{7.86} = 7.12 \text{ CC}$$

$$\frac{7.12}{6.023 \text{ EE23}} = 1.18 \text{ EE-23}$$

B

5

$$\left(\frac{1}{2}\right)^5 = .03+$$

$$\left(\frac{1}{2}\right)^6 = .016+$$

$$\left(\frac{1}{2}\right)^7 = .0078$$

A

6 THE WEIGHT REMAINS 13. ONE PROTON DECAYS TO A NEUTRON BY POSITRON EMISSION.

C

7

$$\frac{(2)(\pi)(.80)}{6} = .838 \text{ Å}$$

E

8

$$M_0 = 9.11 \text{ EE-31 kg}$$

$$k = \frac{1}{\sqrt{1-(.85)^2}} = 1.9$$

$$M_v = (9.11 \text{ EE-31})(1.9) = 1.73 \text{ EE-30}$$

$$E = MC^2 = (1.73 \text{ EE-30})(3 \text{ EE 8})^2$$
$$= 1.56 \text{ EE-13 J}$$

B

9

E

10

E

11

D

12 AFTER 15 HOURS 50% REMAINS.
AFTER 15 MORE HOURS, 25% REMAINS

D

13

C

14

D

15 $\lambda = .693/5700 = .0001216$

$.4 = e^{-.0001216t}$

D

t = 7535 YRS